RESEARCHES ON IRRITABILITY OF PLANTS

RESEARCHES ON IRRITABILITY OF PLANTS

By

JAGADIS CHUNDER BOSE
M.A. D.Sc. C.S.I.

GYAN

Published by

GYAN

Gyan Publishing House
5, Ansari Road
Daryaganj, New Delhi-110002
Phone: 011-47034999, 9811692060
E-mail: books@gyanbooks.com

Distribution Network
gyanbooks.com
India, USA, Canada, UK, Australia, France

ISBN : 978-81-212-6573-7 (PB)
First Published, 1913

2nd Impression 2022

Printed at: Gyan Press, Delhi.

RESEARCHES ON IRRITABILITY OF PLANTS
Author: JAGADIS CHUNDER BOSE

RESEARCHES ON IRRITABILITY OF PLANTS

BY

JAGADIS CHUNDER BOSE, M.A. D.Sc. C.S.I.

PROFESSOR, PRESIDENCY COLLEGE, CALCUTTA

WITH ILLUSTRATIONS

1913

DEDICATED TO

S. M.

PREFACE

I HAVE in this work dealing with my researches on the irritability of plants introduced new methods by which the scope of investigation has been enlarged, and a very high degree of accuracy secured. In my previous treatise on Plant Response, the response recorder employed was a modification of the optical lever, automatic records being secured by the very inconvenient and tedious process of photography. The delay thus imposed retarded seriously the progress of the research. Those practically engaged in investigations on plants can realise the difficulties that arise from the too quick passage of the seasons. It thus frequently happened that by the time new instrumental appliances were rendered practicable the favourable season for the plant was over, involving the postponement of the experiment for another year. In spite of these difficulties, the long series of investigations that I then carried out gave many interesting results, which not only threw light on many obscure problems, but also led to the discovery of several important phenomena in plant physiology.

Some of these results, moreover, tended to cast doubt on certain conclusions that had found universal acceptance. It has, for example, been held that there was no transmission of true excitation in *Mimosa*, the propagated impulse being regarded as merely hydro-mechanical. The question whether the transmitted impulse was physical or physiological could only be satisfactorily decided if the plant could itself be made to record the velocity of its impulse and the changes induced in that velocity under physiological variations. This is but one out of several ideal methods of attacking problems in the life of plants, the realisation of which would make a great advance in physiological investigation.

It would also be desirable to discard, if possible, the troublesome method of obtaining record by photography, which necessitates work in a dark room ; in this connection it should be remembered that subjection of the plant to darkness introduces complications by modifying its normal excitability. For these reasons, another requirement which it is necessary to fulfil is the devising of some simple and direct method of obtaining the record. And in order that the results obtained should not be influenced by any personal factor, it would be further desirable that the plant attached to the recording apparatus should be automatically excited by stimulus absolutely constant, should make its own responsive record, going through its own period of recovery, and embarking on the same cycle over again without assistance at any point on the part of the observer.

The difficulties encountered in realising these ideal requirements appeared at first to be insurmountable. In the records of response serious errors occurred as regards amplitude and time-relations, owing to the friction of the writing lever against the recording surface. As an extreme instance of this, in recording the rhythmic movement of the leaflets of *Desmodium* the very slight friction which the smoked-glass surface offered was enough to stop the pulse-record.

After many attempts, I was at last successful in overcoming all obstacles by the device of the Resonant and Oscillating Recorders. Taking the very difficult test of direct record of the rhythmic movements of *Desmodium* leaflets, it will be found that the pulsations recorded in this book not only gave accurate measure of the amplitude and period, but also the absolute rate of movement during any phase of their autonomous response. Again, in the matter of accurate measurement of short intervals of time required for the determination of the latent period and velocity of transmission of excitation, I have shown the possibility of recording time-intervals as short as a thousandth part of a second. A brief account of this is given in my paper "On an Automatic Method for the Investigation of the Velocity of Transmission of Excitation in *Mimosa*," read before the Royal Society. It will be recognised immediately in how many directions our power of inquiry has become extended by the elaboration of these new methods and the invention of several types of instrumental appliances described in this work.

In presenting the results of these investigations, it will be noted that the plant has been made to tell its own story, by means of its self-made records. Each experiment has been repeated at least a dozen times, in many cases as often as a hundred times. The results may therefore be accepted as fully attested. The establishment of the unity of responsive reactions in the plant and animal, which is the subject of this work, will be found highly significant, since it is only by the study of the simpler phenomena of irritability in the vegetal organisms that we can ever expect to elucidate the more complex physiological reactions in the animal tissues.

I take this opportunity to thank my research assistants, Messrs. Guruprasanna Das, L.M.S., and Surendra Chandra Das, M.A., for the very efficient help rendered by them in these researches.

J. C. BOSE.

PRESIDENCY COLLEGE, CALCUTTA,
October 1912.

CONTENTS

CHAPTER I

PLANT SCRIPTS

CHAPTER II

THE RESONANT RECORDER

CHAPTER III

METHODS OF STIMULATION

CHAPTER IV

TIME-RELATIONS OF THE RESPONSIVE MOVEMENT AND STANDARDISATION OF STIMULUS

CHAPTER V

THE ADDITIVE EFFECT; INFLUENCE OF LOAD, TEMPERATURE, AND INTENSITY OF STIMULUS

CHAPTER VI

VARIOUS TYPES OF RESPONSE

CHAPTER VII

EFFECTS OF DIFFERENT GASES ON EXCITABILITY OF *MIMOSA*

CHAPTER VIII

DEATH-SPASM IN PLANTS

CHAPTER IX

DETERMINATION OF THE LATENT PERIOD

CHAPTER X

INFLUENCE OF INTENSITY OF STIMULUS, FATIGUE, AND TEMPERATURE ON THE LATENT PERIOD

CHAPTER XI

VELOCITY OF TRANSMITTED IMPULSE IN PLANTS

CHAPTER XII

EXCITATORY CHARACTER OF TRANSMITTED IMPULSE IN PLANTS

CHAPTER XIII

THE POSITIVE RESPONSE

CHAPTER XIV

POLAR EFFECTS OF ELECTRICAL CURRENT IN EXCITATION OF PLANTS

CHAPTER XV

POLAR EFFECTS OF FEEBLE AND MODERATE CURRENTS ON VARIOUS SENSITIVE PLANTS

CHAPTER XVI

THE CONTRASTED EFFECTS OF ANODE AND KATHODE

CHAPTER XVII

EFFECT OF TEMPERATURE ON POLAR EXCITATION, AND MULTIPLE EXCITATION UNDER CONSTANT CURRENT

CHAPTER XVIII

POLAR EFFECTS UNDER STRONG CURRENTS

CHAPTER XIX

VARIATION OF POLAR REACTION UNDER TISSUE MODIFICATION

CHAPTER XX

MULTIPLE AND AUTOMATIC RESPONSE

CHAPTER XXI

THE AUTOMATIC PULSATIONS OF *DESMODIUM GYRANS*

CHAPTER XXII

EFFECT OF HYDROSTATIC PRESSURE, LOAD, AND LIGATURE ON THE PULSATION OF *DESMODIUM*

CHAPTER XXIII

EFFECT OF STIMULUS ON LEAFLET OF *DESMODIUM* AT STANDSTILL

CHAPTER XXIV

EFFECT OF ELECTRIC STIMULATION ON THE PULSATION OF *DESMODIUM GYRANS*

CHAPTER XXV

EFFECT OF TEMPERATURE ON RHYTHMIC PULSATION OF *DESMODIUM GYRANS*

CHAPTER XXVI

EFFECT OF CHEMICAL AGENTS ON THE AUTOMATIC PULSATION OF *DESMODIUM GYRANS*

CHAPTER XXVII

ILLUSTRATIONS

RESEARCHES ON IRRITABILITY OF PLANTS

CHAPTER I

PLANT SCRIPTS

Action of environment on plant—Revelation of internal condition by character of response—Problems to be solved—Electrical response—Mechanical response—Motile organ in *Mimosa pudica*—Response in plant and animal—Different phases of the responsive movement—Graphic record—Determination of absolute movement of leaf and its time-relations — Characteristic effects of different agencies on the response-curve — Specific difficulties in recording plant-response.

In strong contrast to the energetic animal, with its various reflex movements and pulsating organs, stands the plant in its apparent placidity and immobility. Yet that same environment, which with its changing influences so strikingly affects the animal, is playing upon it also. Storm and sunshine, the warmth of summer and the frost of winter, drought and rain, all these and many more come and go about it. What coercion do they exercise upon it? What subtle impress do they leave behind? That they, in their totality, do leave the plant better or worse for their occurrence, we know. It is evident that internal changes are effected by their agency which are entirely beyond our visual scrutiny. Would it be possible to trace this general action of the environment into some detail, and then follow out the question of its particular effects upon the vegetal organism? Is there any means by which we might find out whether a given influence has contributed to the plant's well-being or the reverse, whether it has left it more or less excitable, whether it has rendered it more or less energetic?

It is conceivable that internal changes which eluded our direct vision might nevertheless be brought within the range of our observation if we could obtain any sort of answer from the plant itself to a questioning shock. In such a case, the feebleness or vigour of the reply would in itself doubtless constitute a measure of the vitality of the organism. It appears obvious that if any given influence had rendered the plant more excitable, this fact would be manifested by the greater intensity of its response. In a very excitable condition we may suppose the slightest shock of stimulus would evoke a very large responsive expression; in a state of depression, on the other hand, the strongest stimulus would induce only a feeble reply. The relation between the stimulus and the response would thus form a gauge of the physiological condition of the organism. The invisible fluctuating changes taking place in the plant, under the changing conditions of the environment, might in this way be made to reveal themselves.

All this presupposes, however, that the plant will answer in some tangible way to the impinging testing stimulus, and that it may be possible to obtain some record of this answer. The possibility of this will be further discussed presently. There are many important problems which wait for their solution till some such means of inquiry is found. What, for instance, are the various forms of stimulus which evoke an answering reaction in the plant? Again, has a given plant-tissue, like animal muscle, any definite perception-period capable of exact measurement? Is the responding tissue susceptible of fatigue? Is the intensity of its answer dependent on the intensity of the blow? Is the excitation that may be caused at one point transmissible to a distance, as along animal nerve? Is such transmission, supposing it to occur, fundamentally of the same nature as that in the animal? Is there to be found in plants any tissue that might twitch persistently, like the cardiac tissue of the animal? If so, are these rhythmic pulsations characteristically similar? Is there, again, any general resemblance between responsive actions in plant and animal? Going deeper, since the same protoplasmic basis underlies them both, are these reactions to be regarded as essentially the same, though different in degree? If this last were true, then since the simpler explains the more complex, might not the physiological reactions of the plant be expected to elucidate many of the obscurities in the similar reactions of animal tissues?

We return, then, to the question, Is the plant capable of furnishing any such responsive indications as we have supposed? I have shown elsewhere that all plants give response to impinging stimulus by a definite electrical change,[1] which can be recorded by means of suitable apparatus. For the purpose of the present work, however, it will be convenient to employ the more conspicuous motile indications afforded by certain plants, pre-eminent amongst which is *Mimosa pudica*.

The most prominent motile organ in *Mimosa* consists of a mass of tissue known as the pulvinus, at the joint or articulation of the primary leaf-stalk. The swollen mass on the lower side of this organ is very conspicuous. Under excitation the parenchyma, in this more effective lower half, undergoes 'contraction,' in consequence of which there is a fall of the leaf. This sudden movement constitutes the mechanical response of the leaf to the impinging stimulus, just as the contractile movement of a muscle in similar circumstances forms its characteristic mechanical response.

Digressing for a moment to consider the phenomenon of excitatory contractions in general, it may be said that our present knowledge is not complete as to the *minutiæ*

[1] Bose: Friday Evening Discourse—Royal Institution, May 1901; *Comparative Electro-Physiology*, Longman's, London, 1907.

of the process by which these are brought about. In muscle it is supposed that during the act of contraction there is a transfer and redistribution of fluid material.[1] In the case of *Mimosa* there is known to be an escape of fluid from the excited cells; there is a diminution of turgor. It is supposed that this may in some unknown way be connected with a diminution of pressure within the cell.[2]

In the case of the stamens of *Cynereæ*, Pfeffer[3] observed a contraction under excitation of as much as 30 per cent. of the original length. There is an escape of water from the cells into intercellular spaces. The mode in which the fall of turgor takes place is uncertain, and various suppositions have been made to account for it. It has been thought that the escape of fluid is brought about by the elastic cell wall which forces liquid out of the cell, when the protoplasm lining it has become permeable under excitation. There may in addition be an active contraction of protoplasm which might force the liquid out of the cell-vacuole. This latter supposition is regarded by many as improbable, though the observations of Schütt and Benecke indicate that under stimulation the protoplasm of a diatom contracts away from the cell wall. Similar withdrawal of protoplasm

[1] 'Schäfer, working on the highly differentiated wing-muscle of the wasp, concludes that each sarcomere contains a darker substance near the centre, divided into two parts by Hensen's disc. At each end of the sarcomere the contents are clear and hyaline. In the act of contraction the clear material flows, according to Schäfer, into tubular pores, in the central dark material.'—Starling: *Elements of Human Physiology*, 8th edition, p. 91.

[2] 'When the pressure in the cell decreases, we naturally assume this to be due to decreasing osmotic pressure, a decrease which may well amount to 2½ to 5 atmospheres, and may be due either to the transformation of osmotically active substances into bodies with larger molecules, or to alterations in the permeability of the plasma, and an excretion of materials from the cell. As evidence of *excretion of material* we may quote the fact that Pfeffer observed crystals of unknown nature appearing on evaporation of the liquid expressed from the intercellular spaces. Still there are several reasons for doubting this conclusion. It is a remarkable fact that plasmolytic research affords no evidence of any decrease in osmotic pressure."—Jost: *Plant Physiology*, English edition, 1907, p. 515.

[3] *Cf.* Pfeffer: *Physiology of Plants*, vol. iii., English edition, p. 75.

has been observed in *Spirogyra* by Nägeli and in *Nitella* by Hofmeister.

Whatever theory may be held, the undoubted fact in these two cases, of plant and animal alike, is the occurrence of a fundamental excitatory protoplasmic change which finds external expression in alteration of form. If we now record the responsive movement, we shall be recording what is an effect of excitatory change, either in plant or animal. Whether or not this fundamental change is similar in the two cases can only be decided by comparing the records due to excitation, in plant and animal tissues, under all possible variations of external conditions.

The fall of the leaf of *Mimosa* is brought about in consequence of the contraction of cells in the lower half of the pulvinus. I shall for convenience describe the fall as the contractile movement, in contradistinction to the erectile movement brought about by the recovery of cells into normal turgid and expanded condition.

In studying the excitatory reactions of the plant, under external stimulus, we have to determine, first, what time elapses between the incidence of the shock and the initiation of a perceptive responsive movement. This constitutes the determination of the *Latent Period.* We have next to find out at what rate this responsive movement of the leaf takes place, and after what time the contractile phase of the movement is exhausted. After a short pause the plant gradually recovers from the effect of the shock, and the leaf is re-erected to its former position. We therefore want to know the various rates at which recovery gradually takes place. In order to secure these data, it will be necessary to make a graphic record of the entire responsive movement of the plant organ. This record, further, must furnish us not only with the amount, but also with the time-relations, of this movement. This would involve the construction of a writing-lever which, deflected by the pull of the falling leaf, would be capable of tracing on a writing-surface, moving at a known uniform rate, the concomitant curve. For this there must be an axis, supported on frictionless jewelled bearings, and carrying two arms of a horizontal lever and a thin vertical wire with a bent tip, to serve as the writer. The different parts, as far as possible, should be made of aluminium, to secure the utmost lightness. A point of the petiole of the responding leaf would be attached by a silk thread to one arm of the lever, the other having on it a small weight, to act as counterpoise. On the fall of the leaf, under excitation, it would pull down with it the attached arm of the lever. The vertical writer would then also move, say, to the left. If the finely pointed bent end of the writer were to press lightly against the smoked surface of a glass plate, which was allowed to fall, at a uniform rate, by means of clockwork, a curve would then be traced which would not only record the responsive movement and recovery but also give their time-relations (fig. 1). To obtain the latter, it would be necessary to know the rate of movement of the plate on which successive vertical lines might be traced by a time-marker at intervals of, say, one minute.

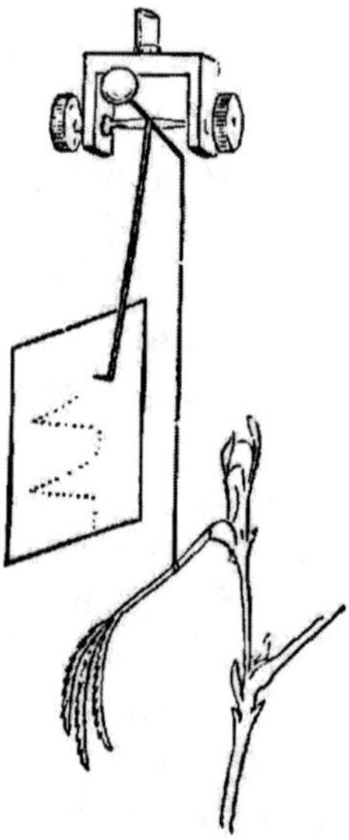

FIG. 1.—Diagrammatic representation of Response Recorder.

In order to find out the absolute movement of the leaf we must know the degree of magnification or reduction that has been effected by the recording arrangement. This will depend upon the relative lengths of the writer and the lever, and the distance of the point of attachment on the leaf from the pulvinus. When the lengths of the lever-arm and the writer are equal, then the writer will describe a movement which is equal to that of the point of leaf-attachment. By shortening the arm of the lever to half the length of the writer, we should obtain the magnification of two. This shortening might be accomplished by attaching the thread nearer to the fulcrum. Proceeding in the manner indicated, any magnification, however high, can be obtained.

Reduction, again, may be effected with equal ease. When the point of attachment is exactly midway between the pulvinus and the tip of the leaf, then the movement executed by it will be half that described by the extremity of the leaf. By bringing the point of attachment nearer

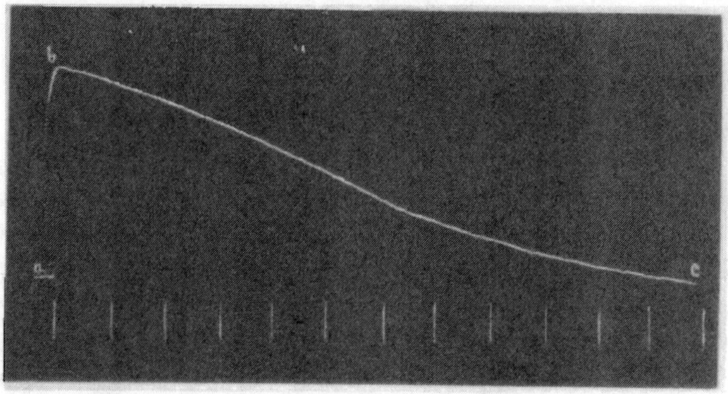

FIG. 2.—Response-curve of primary leaf of *Mimosa*; the vertical lines below the record indicate intervals of one minute each.

to the pulvinus, we can obtain whatever reduction may be required.

The resultant magnification or reduction of the record will thus depend, in any given case, on two factors—namely, the relation between the length of the writer and the length of the lever-arm, on the one hand; and, on the other, the relation between the distance of the point of attachment from the pulvinus and the entire length of the leaf.

Thus if the lever produce a magnification of four, and the point of attachment cause a reduction to half, the resulting magnification will be $4 \times \frac{1}{2} = 2$. As the movement of the primary petiole of *Mimosa* is considerable, the records taken are normally either equal or reduced to two-thirds. A record taken in this manner is given in fig. 2. The height of this curve gives the amplitude of the movement, and the horizontal distance measures the corresponding time. The up-line here, *a b*, indicates the responsive fall, and the descending line, *b c*, the gradual erection, due to recovery. The responsive movement was initiated within an exceedingly short time after the application of stimulus—in this case an electrical shock—and the fall was completed also within a relatively short period. In a record which will be given later, taken on a faster-moving plate, these characteristics will be seen better. The recovery, however, is a slow process, the earlier part being comparatively quick and becoming slower towards the end. The entire recovery is here seen to require 12 minutes.

If we were able to apply a stimulus of exactly identical intensity at regular and suitable intervals, and if the physiological condition of the responding tissue remained constant, then we should obtain a series of responsive twitches which would be practically identical. But if the physiological condition were to undergo any change, under environmental conditions, then the record would give us indications of that internal change, otherwise entirely beyond our power of scrutiny. Thus if the plant were to become depressed, the amplitude of the pulse would undergo a diminution. If on the other hand its excitability should be enhanced, that fact would be indicated by an increase in the amplitude of the response.

The mere amplitude of the twitch, however, affords only a broad indication of the physiological condition of the tissue. There are many factors the effects of which find expression in subtler changes of the response-curve. One agency, for instance, will make the plant more alert. This is at once reflected at that part of the curve which corresponds to the Latent Period. This becomes shorter. The ascent of the curve will also be more abrupt. Another agency will induce, let us say, a contrary change. Different agencies, similarly, will bring about definite changes in the contracting and relaxing portions of the curve. The varying effects of freshness and fatigue, of stimulating or depressing drugs, of heat or cold, of the environment, are in this way clearly revealed by the characteristic flexures of the curve. Thus, by means of testing-blows, we are able to make the plant itself describe those obscure internal changes which would otherwise have entirely escaped us.

In fact, the phytographic records would, in the case of plants, supply us with all the information that myograms afford in the case of animal tissues. The experimental difficulties which the plant offers are, however, very great. In the case of muscle-contraction, the pull exerted is considerable and the friction offered by the recording-surface constitutes no essential difficulty, though even here the time-relations of the curve are, I have reason to think, rendered somewhat unreliable by this friction. In the case of plants the contractile movement is relatively feeble, and in the movement of the leaflet of *Desmodium*, for instance, a weight so small as four-hundredths of a gram is enough to arrest the pulsating leaflets. When employing the very lightest lever, the extremely minute friction offered by the smoked-glass surface of the recording-plate is sufficient in this case to cause complete cessation of the record. Even in the leaf of *Mimosa*, the friction offered is enough to distort the curve to such a serious extent that errors are introduced into the amplitude and time-relations amounting to more than 50 per cent. These difficulties have been overcome by the successful devising of my Resonant Recorder, an account of which will form the subject of the next chapter.

Summary

Obscure modifications of internal condition of plant, resulting from changing factors of environment, may be revealed by records of plant's response to testing-blows.

The relation between the impinging stimulus and the

intensity of reply measures the physiological efficiency of the plant tissue for the time being.

Automatic records of mechanical and electrical responses of plant can be obtained with suitable apparatus.

For recording the mechanical responses of *Mimosa* and other sensitive plants, a writing-lever is employed. Distortion of record by friction is apt to introduce error, which has to be eliminated.

CHAPTER II

THE RESONANT RECORDER

Advantages of intermittent contact in record—Two types of apparatus: the Oscillating Recorder and the Resonant Recorder—Coercer and Vibrator—Perfect tuning—Recorders with standardised frequencies—Slide and clockwork—The record its own chronogram—Smoked surface and its fixation—Adjustments of the writer—Records with continuous and intermittent contacts.

It was stated in the last chapter that however light the contact may be, and however smooth the glass recording-surface, the record was still apt to be either arrested, or seriously distorted, on account of friction. As long as I employed the ordinary method of continuous contact of the writing-point with the glass surface, it was impossible to overcome this particular difficulty. It occurred to me at last that the problem might find a solution if I could succeed in making an intermittent instead of a continuous writing-contact. I have solved this problem by devising two different types of apparatus, which I have called respectively the Oscillating Recorder and the Resonant Recorder. In the former, the recording-surface itself is made by an electro-magnetic device, to vibrate to and fro, thus bringing it into periodic contact with the writing-point.[1] This apparatus is extremely convenient for the general purpose of recording responses in which the measurement of excessively short intervals of time is not essential.

But there are many important problems, such as the determination of the latent period, and accurate determination of the velocity of transmission of excitation, in which

[1] Bose: *British Association Report*, Dublin, 1908, p. 903.

we require time-measurements of the order of one-hundredth of a second. It will be shown that these determinations can be carried out with great precision, by means of the intermittent dots themselves, when the periodicity of their recurrence is rendered perfectly constant. For such purposes, then, we require a frequency of intermittent contacts amounting to something like a hundred times per second.

It was clearly impossible to make the heavy plate-carrier oscillate with such a high frequency. There remained only the theoretical alternative of causing the writing-point to vibrate to and fro, at the required frequency, so as to make the necessary intermittent contacts with the surface of the recording-plate.

The advantage of this intermittence may be understood from a concrete example. It will be remembered that the writing-point, under the action of the responsive fall, moves parallel to the surface of the recording-plate. If now, by means of some mechanism, the writing-point be made to vibrate to and fro, say, ten times each second, at right angles to the plate, this will in no way affect the record beyond the fact that instead of a continuous line a dotted line will be traced. The record will not now labour under the defects inseparable from the friction of continuous contact. Instead of this, we shall have the vibrating writer tapping a record which is practically free from friction. For it will be understood that, as in our concrete example, a recording-point which is vibrating ten times each second will execute one entire to-and-fro movement in one-tenth of a second. The duration of contact, at the extreme forward end of the swing, will represent only a small fraction, say one-fifth, of the entire period of one vibration. Hence after each contact, lasting only one-fiftieth of a second, the recording-point is absolutely free to take up the movement impressed upon it by the moving leaf. In a record lasting for one second the sum of the intermittent contacts will then amount to one-fifth, and the period of entire freedom to four-fifths of a second. We can thus see the theoretical advantages of an intermittent over a continuous contact. What has been said of the writer vibrating ten times in a second holds good equally in those cases where the vibration-frequency is much higher.

One great difficulty we encounter in carrying out this idea lies in giving the recording-point an impulse exactly perpendicular to the direction of its recording movement. If the recording-writer be made of fine steel-wire, and if we place behind it a small electro-magnet—the pole consisting of a rectangular piece of soft iron, at right angles to the wire—then, by sending a momentary strong current through this electro-magnet, a pull will be exerted on the wire which will make its recording-tip strike for an instant against the glass recording-surface. As the steel wire has to be made extremely fine, in order to reduce to a minimum the inertia of the recorder, the resulting pull exerted on it is very slight, unless an excessively strong current be sent round the electro-magnet. Again, unless the intermittent closures of the electric circuit be properly timed, the writing-index may be subjected to attraction in the course of its journey, now to the recording-surface and again away from it. In the latter of these cases its vibration, on which the intermittent contact depends, is totally destroyed.

But the most serious difficulty of all is that introduced by the edge of the attracting electro-magnet. It is known that the magnetic intensity of a pole is strongest at its edges. Should the writing-index by chance be placed exactly symmetrically, as regards the right and left edges of the pole, then the two lateral pulls, being equal, will neutralise each other, and the index will vibrate to and fro perpendicularly to the recording-surface. But should it be placed, however slightly, nearer to one edge than to the other, then one of the two pulls will be in excess, and the index will be drawn to one side, thus producing a disturbance in the record not due to the excitatory pull of the leaf. Even if, at the beginning, the index had been placed in a strictly symmetrical position, the movement of the writer caused by the excitation of the leaf would draw it into an asymmetrical position, resulting in a one-sided pull which would seriously interfere with the reliability of the record.

The Resonant Recorder

It is therefore absolutely necessary so to arrange matters that the electro-magnet shall be without laterality. This condition I was able to fulfil by making the pole of the electro-magnet in the form of either a cylinder or a ring. The axis, from which is suspended the writing-index, is accurately supported, perpendicular to the plane of the circular section of the magnetic pole at its centre. Everything was thus made symmetrical, and as there was no laterality there could be no tendency whatsoever for the index to execute its to-and-fro vibrations in any other direction than that which was perpendicular to the plane of the terminal pole of the magnet. As this plane may be adjusted parallel to the glass recording-surface, the tapping movement of the writing-index can be made to take place perpendicularly to the recording-surface.

Next, in order to overcome the difficulty of the irregular timing of those electrical impulses which are to maintain the recording-index or writer in a state of periodic vibration, I devised the Resonant Recorder. If we know the natural frequency of vibration of the recording-index, and if by means of some mechanism we can send periodic currents of exactly the same frequency through the electro-magnet, then the intermittent magnetic pulls will exactly synchronise with the natural swings of the writing-index. Owing to this perfect tuning the index will now resonate, breaking out into a persistent and regular vibration of considerable amplitude. In practice, all that is necessary in order to secure this is to take a long steel reed, which in the course of its regular vibration will periodically interrupt the electro-magnet circuit of the vibrator coil. The reed itself is maintained in a state of persistent vibration by the usual electro-magnetic arrangement. I shall for the sake of convenience refer to this reed-interrupter as the Coercer ; and the writing-index simply as the vibrating-recorder or Vibrator. The reed was at first purposely selected of too great a length, so that the natural frequency of the coercer should be slower than that of the vibrator. The free end of the coercing reed carries a platinum wire which, dipping into a cup of mercury, completes the electric circuit. The other end is clamped, and by shifting the clamping-point the vibrating length of the reed is gradually and continuously shortened. This has the effect of gradually raising the vibration-frequency of the interrupting reed. A time soon comes when the frequency of the coercer is exactly the same as that of the vibrator. The latter, which has been hitherto more or less inert, now suddenly breaks out, as foreseen, into very regular and sustained vibrations of large amplitude. For some purposes it is important that the vibration-frequency of the recording-index should have a definite value. The various frequencies most suitable for these researches were 10, 20, 50, 100, and 200 vibrations per second. The exciting reed was previously calibrated by means of frequency-meters, or standard tuning-forks, to give these values. Then, with great expenditure of time and patience, different vibrating-recorders having various standardised frequencies were constructed. For this we have to select fine-steel wires of differing lengths and thicknesses. The final tuning is accomplished by careful filing or hammering. Filing the tip of the vibrator raises the frequency ; and hammering of the wire, near the point of suspension, lowers it. A general flattening, along the whole length, tends to maintain the vibration in a definite plane, otherwise the free tip is apt to execute an elliptical vibration. The tuning of the vibrator with the coercer is not very difficult when the vibration-frequency is low, say 10 per second. But when the frequency is high, say 100 or 200 times in a second, an exact tuning is essential. The slightest variation of the length of the coercer will either bring about full resonance or make it entirely ineffective. When the tuning is nearly but not quite perfect, then we have the phenomenon of beats. In this case, in the successive dots of the record, there will be periodic blanks. For the purpose of exact adjustment of length of the coercer I employ a micrometer-screw, by means of which the most delicate adjustment of length may be carried out.

For periodic interruption the coercing and vibrating coils may be put in series, but I find it is much easier to obtain a persistent vibration when the coercer coil is placed in a multiple arc with the vibrator coil. An electro-motive force of 4 volts should be sufficient for the purpose of maintaining a steady vibration of both the coercer and the vibrator.

Having thus secured the requisite perfection of the resonating-writer, it is necessary to describe the complete apparatus by which to obtain records of responses in *Mimosa* and other sensitive plants. For this purpose we require a slide-carrier to hold the recording-plate, and this is to be dropped at a definite speed, without jar ; also the clockwork by which it is to be actuated. Besides these is needed some special means by which the recording-point may be brought to the proper distance from the recording-surface. It is necessary, again, that the response-movement of the writer should be absolutely parallel to the writing-surface, and that its tip or contact-point should be capable of delicate adjustment as regards distance. It should be possible, moreover, to bring this writing-point to any position on the recording-surface that may be required. I will now proceed to relate the devices by means of which all these conditions have been met. Some of these will be seen in fig. 3, which illustrates only the upper part of the Resonant Recorder.

The Slide and Clockwork

A gunmetal upright, the upper part of which is of triangular section, stands on a large disc of the same metal, which is screwed to a larger wooden base-board. The slide-carrier, holding the glass recording-plate, moves up and down the top part of the upright. It is essential to have this slide so accurately fitted that the plate-carrier may be able to drop

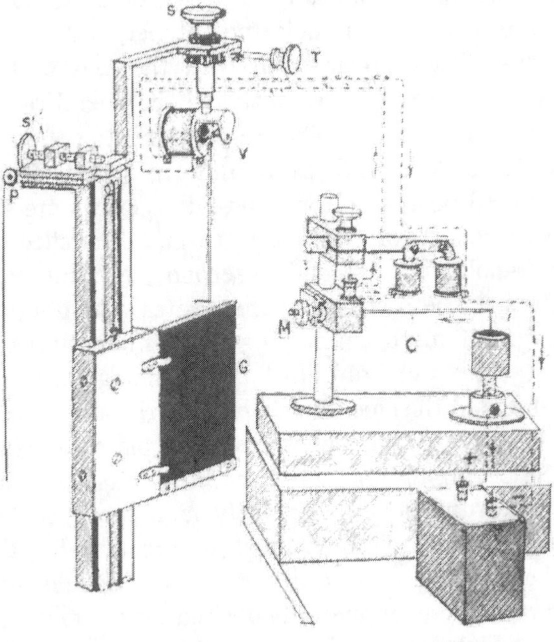

FIG. 3.—Upper part of Resonant Recorder. (From a Photograph.)

Thread from clock, not shown, passes over pulley P, letting down recording-plate; S′, screw for adjusting distance of writing-point from plate; S, screw for vertical adjustment; T, tangent-screw for exact adjustment of plane of movement of recorder, parallel to writing-surface; axis of writer supported perpendicularly at centre of circular end of magnet; C, coercer; M, micrometer-screw for adjustment of length of coercer; V, vibrating recorder; G, smoked-glass plate.

smoothly and uniformly, without any jerking whatsoever. I have sometimes attained the same end by mounting the plate-carrier on wheels and letting it slide down vertical rails. The plate-carrier is allowed to drop by the running down of a clock or a phonograph motor, according to what may be the requirement of the speed. The quarter-plate size (11 × 8 cm.) is convenient for record, as it is not too large for book illustration. The suspending thread passing over pulleys is wound round the winding-wheel of the clock. This wheel is provided with click and ratchet, which allow it to be wound without interfering with the axis of the clock. Thus winding of the wheel in the left-handed direction pulls up the recording-slide. The running-down of the clock then allows the slide to fall at a uniform rate. The various speeds found necessary for different records were such that the entire length of the plate, 11 cm., travelled past the recording-point in .5 second, 6 seconds, 15 minutes, 1 hour, or 3 hours. The first two of these rates were obtained from a phonograph motor employing two different-sized wheels. The last three were obtained by attaching three different-sized wheels to the clock-axis which carries the minute-hand. In these slow rates the movement of the plate is quite uniform from the beginning, but when, as in the first two cases, this has to be dropped at a relatively high speed, a short time will elapse, equivalent at most to the first fourth of the plate, before it becomes quite uniform. Should uniformity of such movement be specially desired for the record, it must be commenced after passing this first fourth. But on account of the chronographic signals which accompany the record, this uniformity is not absolutely essential, for they give us the data from which the time-relations of the curve may be derived.

I may here refer to a few practical points with regard to the preparation of the glass for record and its subsequent fixing. In order to produce an even layer of smoke on the recording-plate, it is moved over the gas-flame from a bat's-wing burner ; and this deposit of smoke will be improved if the gas has been previously passed through a jar containing a small quantity of benzine. After the record is taken, it is fixed by pouring carefully over it a dilute solution of canada balsam in xylol. It is afterwards easy to reproduce this record by contact-print on a photographic paper.

Adjustment of the Writer

It is sometimes necessary to have the recording-point brought two or more times to the same place, in order that the successive records may be rendered the more strictly comparable. This is accomplished by a rack and pinion to adjust the height of the platform carrying the plant. When the platform is lowered, the petiole, which is attached to one arm of the recording-lever, pulls it down, and the recording-point is moved to the left. When the platform is raised, then by the action of the counterpoise attached to the other arm of the lever the index is moved in the opposite direction. In this way the recording-point can be brought to any position that is desired. The same end is secured through adjustments of a micrometer by which the carrier of the writing-index is raised or lowered.

Another adjustment that is necessary is the bringing of the recording-point near to the writing-surface without actual contact; so that, when the index is set in a state of resonance, it may trace a dotted line. The necessary adjustment is brought about by means of a micrometer-screw at the top of the instrument, by which the lever can be made to approach or recede from the writing-surface. When the speed of the plate is slow, the successive dots may be so close together as to appear like a continuous line.

More troublesome is the adjustment necessary to render the plane of movement of the index exactly parallel to the writing-surface. If this be omitted, the writing-point in one part of the record, say to the right, will be too far away to strike the surface, whereas in another part, say to the left, it will press against the plate and lose its freedom. This difficulty I have been able to overcome by mounting the vertical rod, carrying the writer, inside another tube. An attached tangent-screw, T, then causes a very slow rotation of the vertical rod, either in the right-hand or the left-hand direction. By this means it is possible to bring the

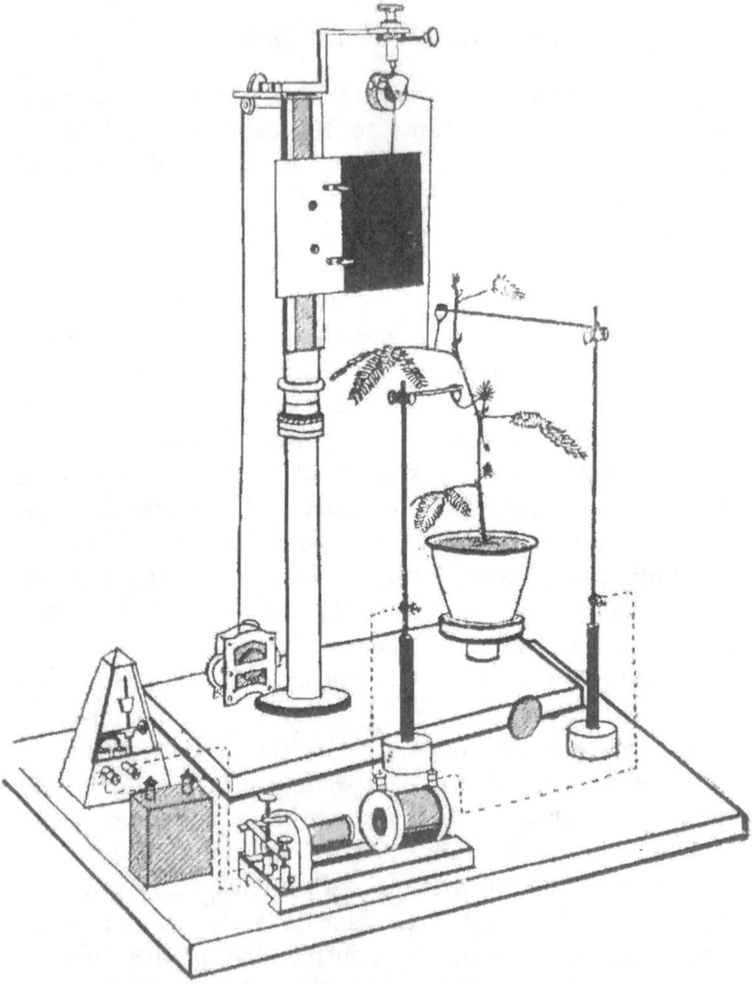

FIG. 4.—General view of the whole apparatus, and the electrical connections by means of which excitatory shock of a definite duration may be given to the plant; duration of shock determined by metronome which completes electrical circuit. (From a Photograph.)

plane of movement of the writing-index exactly parallel to that of the writing-surface. The complete apparatus for obtaining response of *Mimosa* is shown in the accompanying illustration (fig. 4).

Having thus given an account, in some detail, of the practical working of the Resonant Recorder, it will now be well to show a pair of curves which demonstrate, in a marked manner, the advantage of intermittent over continuous contact in the making of these records (fig. 5). These represent two successive experiments on the same leaf, under identical stimulation of an electrical shock. The recording-plate was here moving at a moderately high speed. The lower record was taken with continuous contact, and the upper with the same recorder but in a state of vibration, giving intermittent contact. The vibration-frequency was 10 times per second. Stimulus was applied at the point marked by the vertical line. A comparison of the two records will show that owing to the relative loss of freedom, due to friction, in the continuous contact, the latent period, or the interval between stimulus and initiation of response, is prolonged and the amplitude of the response itself reduced. In the case of the intermittent contact, on the other hand, we see that besides the freedom from this particular error we have the further advantage that the record itself contains its own time-marks, the successive dots being at intervals of one-tenth of a second.

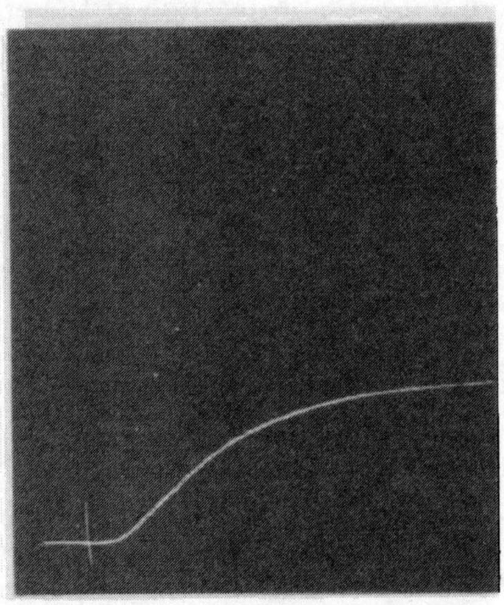

FIG. 5.—Advantage of intermittent over continuous contact in obtaining records.

We have next to consider the practicability of devising,

for the excitation of the plant tissue, perfect methods of stimulation, the intensity of which can either be maintained constant or varied in a perfectly known manner. We have moreover to render the successive stimulations, and consequent scripts of the plant, a perfectly automatic process; so that the experimenter may be comparatively relieved of personal participation in the securing of the records. This will have the incomparable advantage of having no element of personal error in the results so obtained. The question of the effects of the various forms of stimulus will be dealt with in the next chapter.

SUMMARY

In the response-records of plants, errors are introduced on account of friction of the writing-point against the recording-surface.

These errors are eliminated by the method of intermittent instead of continuous contact for the record. By employing the principle of resonance, the writer is made to vibrate to and fro at a known and definite rate. The record consists of series of dots giving definite time-intervals. The record is thus its own chronogram.

CHAPTER III

METHODS OF STIMULATION

Different methods of stimulating the plant: mechanical, chemical, thermal, and electrical—Difficulties of securing quantitative stimuli—Direct and indirect stimulation—Ideal modes of stimulation—Electro-thermic stimulation—Stimulation by constant current—Stimulation by condenser-discharge—Non-polarisable electrodes—Direct, extra-electrodal, and intra-electrodal stimulation—Stimulation by induction shock—Effects of make- and break-shock—Excitation by tetanising shock.

In the case of contractile animal muscle, various stimuli give rise to excitation, and it is a very remarkable fact that the same stimuli exercise a similar excitatory influence on the pulvinus of *Mimosa*. Classifying these stimuli, we find that they are :—

1. *Mechanical.*—A blow will excite animal muscle and cause mechanical response. A similar effect is induced by a mechanical blow in the pulvinus of *Mimosa*. A prick or cut also will cause contraction in either.

2. *Chemical.* — Various chemical agents are found to induce excitation in both animal and vegetal contractile tissues. Thus dilute hydrochloric acid or ammonia causes excitation of both muscle and pulvinus.

3. *Thermal.*—The application of a hot wire will induce responsive contraction in both cases.

4. *Electrical.*—The muscle may be excited by an induction-shock. The pulvinus of *Mimosa* is also excited by such shocks. Other modes of electrical stimulation, such as that of condenser-discharge and that of the application of a constant electrical current, are found effective in causing excitation of animal tissues. It will be seen in the

course of the present chapter that plant tissues also may be excited by similar methods.

In all these cases excitation may be either direct or indirect. In the case of muscle, with its attached nerve, we may cause excitation directly by applying the various forms of stimulus on the muscle itself, or indirectly by applying them on the nerve. In the latter case excitation is transmitted by the conducting-tract—the nerve—and reaching the muscle after a brief and definite interval, induces there the usual contraction.

Taking the case of *Mimosa*, we may similarly have either direct or indirect excitation. Excitation is direct when it is applied, say, on the contractile pulvinus itself. It is indirect when it is applied on the petiole, at a distance from the pulvinus. Certain tissues in the petiole conduct this excitation, which, reaching the pulvinus after a definite interval, induces a responsive contraction.

It is usually maintained that in the case of *Mimosa* there is no true conduction of excitation, but that this contention is not justified will be fully demonstrated in a subsequent chapter. We have, then, in correspondence to the nerve and muscle preparations of the animal, plant-specimens, consisting of petiole and pulvinus. Indirect excitation for specific experiments is effected in the animal through the nerve, and in the plant through conducting-strands embedded in a tissue, as in the petiole.

Although we thus have various forms of stimulus at our disposal for inducing individual and isolated responsive contractions in *Mimosa*, yet we are confronted with very great difficulties when we wish to obtain a series of uniform excitations for quantitative investigation. It is obvious that chemical forms of stimulus would be impossible for successive excitations. The objection to a mechanical blow, as the stimulus to be employed, lies in its liability to cause a mechanical jar and thus to disturb the record.

The ideal form of stimulation would be one the intensity of which might be maintained uniform in successive experiments, or varied in a definite and known manner. Another great obstacle to be overcome in practice is the avoidance of injury which is caused by the stimulus itself. The application of stimulus above a critical intensity induces a depression or abolition of excitability of the tissue.

As the result of long investigation for the purpose of securing various forms of quantitative stimulus, I find that one mode of thermal and three modes of electrical stimulation may be rendered practicable for our purpose. These four different methods will be described in some detail below.

ELECTRO-THERMIC STIMULATION

It is evident that touching the specimen with a hot wire, though effective, is not a form of stimulus that is capable of quantitative application or of repetition. It is apt, moreover, unless very great precautions are taken, to injure the tissue.

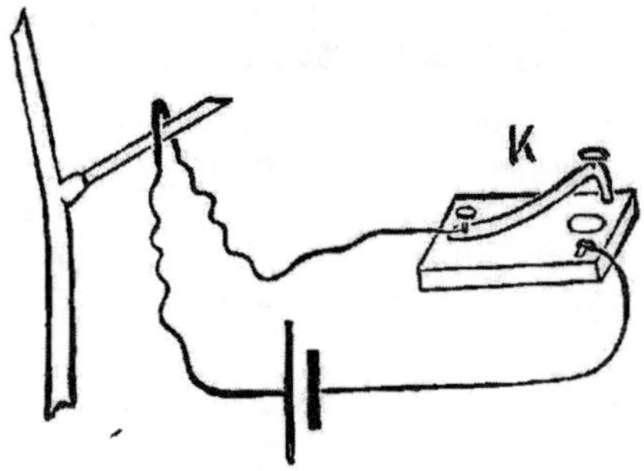

FIG. 6.—Electro-thermic stimulator for uniform stimulation; metronome employed in place of key K, for closing circuit for definite length of time.

The thermal mode of stimulation can, however, be rendered practicable by the electrical mode of the generation of heat. A loop of fine platinum-wire is made to clasp round the petiole which is to be excited, and is connected with an electrical circuit by means of fine flexible silver-wire (fig. 6). The circuit can be completed by a metronome interrupter, the current from the battery flowing for a definite length of time during, say, a single or definite number of beats of the metronome. This produces a sudden thermal shock, enough to cause excitation. Successive uniform stimuli can thus be applied. By means of a variable resistance included in the circuit, the intensity of the stimulus can be increased or diminished. Care must, however, be taken that the heat produced in the platinum loop shall not be such as to scorch or otherwise injure the tissue.

I find that injury from scorching may be avoided by adding a drop of water at the point of contact and afterwards removing excess of water by blotting-paper. This thin film of water protects the tissue from a burn. It is, again, not absolutely necessary to place the platinum wire in contact with the plant. Excitation will take place if the heating-wire is in close proximity. How practicable this form of stimulus may be rendered will be observed from the record (fig. 7) of two successive excitations by this method, which are seen to be uniform. For certain electrical investigations it is essential that stimulus other than electrical should be employed. This requirement is admirably fulfilled by the thermal mode of stimulation.

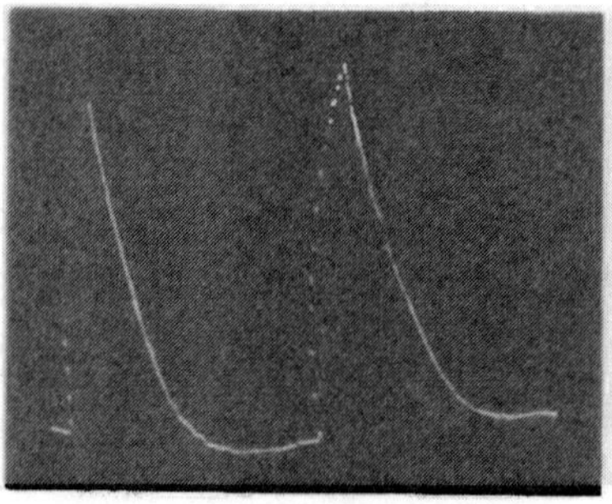

FIG. 7.—Response records of *Mimosa* under indirect thermal stimulation.

Another mode of stimulation—namely, that of thermal radiation—can also be employed, though not so conveniently as the former. A certain area may be rendered radiant by the passage of an electrical current. A Nernst electrical lamp can be conveniently utilised for the purpose. This, when rendered incandescent, gives out radiation of constant intensity. This radiation consists not only of light rays but also of a large proportion of obscure heat-rays. The excitatory value of the latter is more efficient than the luminous rays. The radiant surface of the Nernst lamp is suitably placed in front of a concave metal mirror, by means of which the rays can be focused upon any point that is desired.

Stimulation by Constant Current

I have found that *Mimosa* and other sensitive plants show certain very remarkable excitatory effects under the action of a constant current. The characteristic feature of these is that excitation is not induced during the passage of the current but only at its initiation or cessation. The excitatory effect in this case is further conditioned by the point

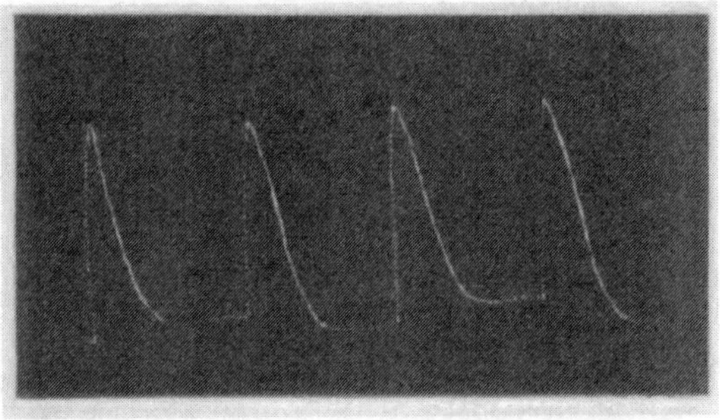

Fig. 8.—Responses to stimulation by constant electric current.

of entry, or anode, and that of exit, or kathode. The specific characteristics of this mode of stimulation will be found fully described in the chapter on the Polar Effects of Currents in Excitation. It need only be mentioned here that, in the matter of all these peculiar effects, the plant tissue behaves in a manner exactly similar to the animal tissue. A series of records obtained from *Mimosa* by the stimulus of a constant current are shown in fig. 8.

Stimulation by Condenser Discharge

Another practical method of stimulation is that of condenser discharge. The condenser consists essentially of two conducting-plates—which may be two sheets of tin-foil—separated by a sheet of non-conducting material, such as mica or paraffined paper. The capacity of the condenser is increased by enlarging the effective area of the plates. The diagram in fig. 9 illustrates this mode of excitation. By increasing the number of cells, the charging E.M.F. may be increased until a suitable value is obtained which is efficient for excitation. This will depend on the excitability of the plant-specimen. About 2 volts charging ·5 microfarad will in general be found sufficient. K is a special spring-key by which the condenser may be charged

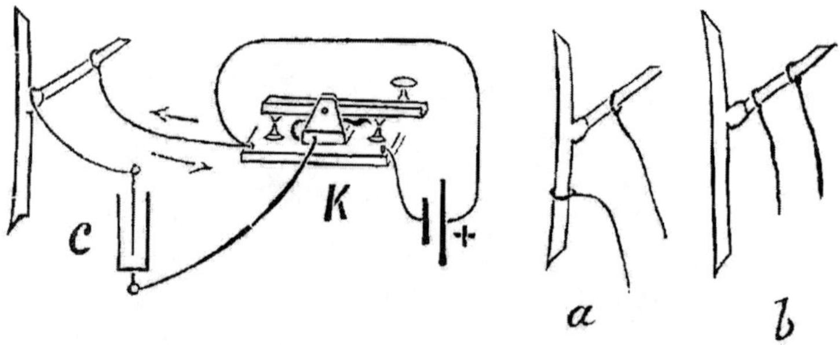

FIG. 9.—Direct stimulation by condenser discharge; C, condenser, K, key. (*a*) intra-electrodal and (*b*) indirect extra-electrodal mode of stimulation.

or discharged. The plant to be excited is included in the electrical circuit. When K is pressed down, the condenser is charged, the instantaneous charging current passing in one direction. The upper arrow in the diagram shows the direction of this charging current. When the key is released, it springs back and discharges the condenser. The instantaneous discharge current now flows in a reverse direction (fig. 9).

The electrical connections with the plant are diagramatically shown, in this and other figures, by two lines. In practice the connections have to be made by means of thread moistened in dilute saline solution. In certain experiments it is necessary to avoid complications arising from electrolytic polarisation. In these it is advisable to use non-polarisable electrodes for making the connections with the plant tissue. Such non-polarisable electrodes may be of the usual U-tube type. The shorter limb of the glass U-tube is filled with kaolin paste in normal saline. A cotton thread moistened in saline protrudes from this and makes the connections with the tissue. The longer limb is filled with zinc sulphate solution, into which dips a zinc rod. For ordinary purposes, however, a much simpler contrivance is found effective. A narrow cork is partially hollowed out and paraffined. The well thus formed is filled with dilute saline solution. The bottom is pierced for the entry of a cotton thread into the saline. A thick silver-wire, whose surface has been covered electrolytically with a film of chloride, pierces the side of the cork and dips into the saline solution. The silver wire forms one of the two electrodes, and the cotton thread makes the necessary electrical connection with the tissue.

For the purpose of excitation we may make the two electrical connections, one at or near the pulvinus itself, and the other on the petiole at a short distance. It will be shown later that when the electrical current leaves the tissue by the pulvinus, that point becomes the seat of excitation. Thus by making the pulvinus the point of exit of current, or kathode, we may cause direct excitation. Or we may have the pulvinus included between the two electrodes, so that the electrical current passes through it (fig. 9, *a*). This connection we may designate the intra-electrodal. Here, in certain circumstances, the excitation throughout the tract becomes diffuse and practically instantaneous. And lastly, the two electrical connections may be made side by side, say about 1 cm. apart on the petiole, at a moderate distance from the pulvinus. The excitation thus caused in the petiole reaches the pulvinus, as I have already said, by conduction. This connection we may call extra-electrodal (fig. 9, *b*).

I give below a series of records (fig. 10) of response to stimulation by condenser discharge. The plant was highly excitable, and excitation was caused by the discharge of ·1 microfarad condenser charged to ·3 volt.

Stimulation by Induction-shock

Excitation may be induced by means of a single or repeated shock from an induction coil. In my own experience I was at first under the impression that this mode of stimulation was not suitable for repeated quantitative experiments, as I found that the plant was liable to become insensitive owing to the fatigue or injury caused by the shock. Later, however, I was able to trace this difficulty to the

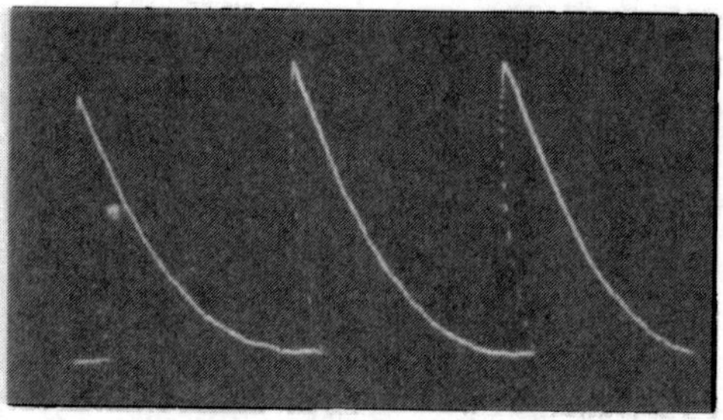

Fig. 10.—Record of responses to stimulation by condenser discharge.

employment of an intensity of shock which was in excess of a certain critical value. I had in fact been misled by the prevailing belief that the excitability of the plant was considerably lower than that of the animal. Hence I employed an intensity of current which was unnecessarily high. This induced fatigue and consequent insensitiveness. Afterwards I discovered that in so far as its sensitiveness to electrical stimulation was concerned, *Mimosa* was in no way inferior to the animal. Quantitative results will be given later, in justification of this statement. Avoiding, then, an intensity of stimulus which was too great, and allowing proper resting-intervals, I found that the efficiency of induction-shock as a mode of stimulation was all that could be desired. It has also the great advantage of allowing successive stimuli to be maintained constant, or to be increased in a known manner.

The induction coil consists of a primary made of a few turns of thick wire enclosing a bundle of soft iron wire—thus forming an electro-magnet—and of a secondary consisting of a larger number of turns of thin wire. The secondary coil can be made to approach or recede from the primary, by means of a slide. When a current is suddenly started in the primary coil, by pressing a key an instantaneous make-induction-current is induced in the secondary. When, by releasing the key, this current is broken, an instantaneous break-induction-current, whose direction is opposite to that of make, is induced in the secondary. Owing to the greater suddenness with which the break is effected, the intensity of the break-shock is greater and of shorter duration than that of the make-shock. The intensity of either make- or break-shock may be increased by bringing the secondary nearer the primary. We can obtain successive uniform shocks, of either make or break, by maintaining the distance between the two coils constant.

We may subject the tissue to shock of either make or break at will by employing an additional short-circuiting key. When this key is down, the shock from the secondary coil is practically diverted across the better conducting-path provided by the key, so that for practical purposes none passes through the plant tissue. If it is desired to cause successive excitations by make-shock only, then the short-circuit key is raised when the primary circuit is made, and pressed down when this is broken. For exciting by break-shock, the short-circuit key is pressed when the primary is made and raised when the primary is broken (fig. 11).

Effects of Make- and Break-shock

In the response of animal tissue it is well known that while single induction-shocks are effective in the case of quickly reacting skeletal muscles, they induce hardly any contractile effect in the more sluggish smooth muscles. Vegetal protoplasm also is commonly regarded as little capable of excitation by these shocks, behaving in this respect like the sluggish smooth muscles amongst animal tissues. Again, while the break-induction-shock of higher intensity and shorter duration is more effective in exciting the quickly reacting skeletal muscle, in the case of the sluggish smooth muscle it is the make-shock of low intensity and long duration that proves more efficacious. That

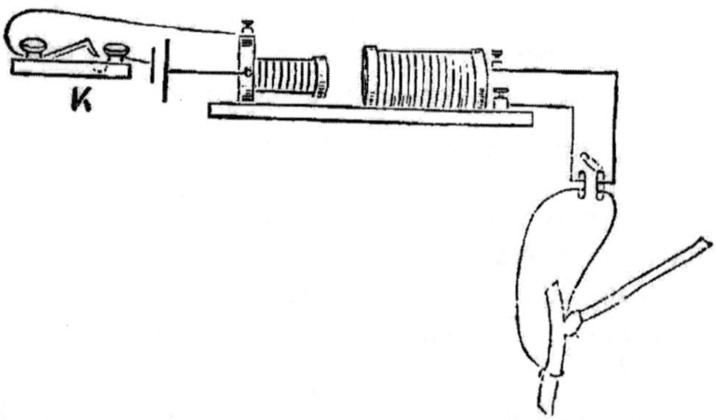

FIG. 11.—Arrangement for applying single make- or break-shock; K, key in the primary circuit. The secondary circuit may be short-circuited by the second key.

the inference commonly made about the reaction of vegetal protoplasm to single induction-shocks is not of universal application, is strikingly seen in the response of pulvinus of *Mimosa*. Here, so far at least as single induction-shocks are concerned, its reaction appears more analogous to that of skeletal than of smooth muscle; as a position of the secondary in relation to the primary can be found in which, while a single make-shock is ineffective, a single break-shock is quite efficient. In order to render the make-shock effective, the secondary has here to be pushed in nearer to the primary, thus increasing the intensity of the shock. A pair of records will be given in a later chapter (figs. 20, 21), showing the relative ineffectiveness of the make-shock compared with the break-shock.

EXCITATION BY TETANISING SHOCKS

It will be shown later that shocks individually ineffective become effective by repetition. It is thus possible to excite a plant by subjecting it to a number of relatively feeble make- and break-shocks. The advantage of this mode of stimulation is that, owing to the low intensity of these shocks, the liability of the tissue to injury is very much reduced. Such alternating tetanising shocks can be produced by means of an automatic spring-interrupter, included in the primary circuits. This interrupter consists of a steel spring carrying at its free end a soft-iron armature which faces one pole of the electro-magnet of the primary. An adjustable contact-rod touches the spring and completes the primary circuit. But the completion of the circuit magnetises the electro-magnet, which, pulling the armature, breaks the contact, thereby interrupting the primary current. The electro-magnet is thus demagnetised, the armature is released, and the spring returns suddenly, re-establishing the circuit. By this automatic make-and-break we obtain alternating induction-currents in the secondary.

When we wish to subject the experimental tissue to the additive effects of these shocks, of a given short duration, this may be accomplished by including in the primary circuit a metronome, which in the course of a single beat closes the circuit for an approximately definite duration. If the duration of closure be, say, one-fifth of a second, and if the frequency of the spring-interrupter be 50 times per second, the number of alternating double-shocks given to the tissue would be ten.

A method of securing still greater accuracy in the duration of the tetanising shock will be described in another chapter, where also may be seen records obtained by that mode of excitation. In subsequent chapters we shall also study in detail the various characteristics of response and its time-relations.

Summary

The ideal method of stimulating a plant is one in which the intensity might be maintained uniform or varied in a definite and known manner.

If the stimulus exceeds a certain critical value, the tissue is injured with concomitant diminution or abolition of excitability.

One practical method of quantitative stimulation is by electro-thermic stimulus, where the plant-tissue is subjected to a sudden and definite thermal variation.

The plant may also be excited by the action of a constant current.

Another method of excitation is by the discharge of a condenser.

And lastly, excitation may be induced in the plant by a single induction-shock or by a series. As in the skeletal muscle of animals, so in the pulvinus of *Mimosa*, the break-shock is more effective than the make-shock.

CHAPTER IV

TIME-RELATIONS OF THE RESPONSIVE MOVEMENT AND STANDARDISATION OF STIMULUS

Latent period of *Mimosa*—Apex-time—Rate of responsive movement of leaf—Effect of intensity of stimulus, fatigue, and temperature—Periodic dot-marker—Time-relations of response and recovery—Effect of season—Response of *Biophytum*—Response of *Neptunia*—Arbitrary distinction between sensitive and ordinary plants—Differential response in *Mimosa*—Response of ordinary plants—Universal sensitiveness of plants—Standardisation of stimulus—Maximal and minimal stimuli—Extreme sensitiveness of *Mimosa*.

As already stated, when the pulvinus of *Mimosa* is subjected to an instantaneous stimulus, say that caused by an electric shock, a responsive movement is initiated after the lapse of a very short interval. After the completion of the fall of the leaf, the contracted pulvinus slowly recovers its original expanded condition, with consequent re-erection of the leaf. The movement of the leaf is thus a visible indication of the responsive reaction and recovery of the pulvinus under stimulus. In this entire process, we may conveniently distinguish three separate phases :—

First, there is a brief period between the incidence of stimulus and beginning of the responsive movement : the contraction has not yet manifested itself. This lost time is called the *Latent Period*.

Secondly, after the lapse of the latent period, the leaf begins to fall, at first with increasing rapidity, which then again diminishes, till it comes to a stop. The curve described attains its maximum amplitude, corresponding to the maximum fall of the leaf. The period required, up to this point, we shall call the *Apex Time*. The pulvinus remains for a short time in its contracted condition.

Third and lastly, recovery of the pulvinus from the effect of stimulus begins to take place, with consequent re-erection of the leaf. This process of recovery is very much slower than the responsive fall. While the responsive fall is a matter of a few seconds only, the re-erection or recovery requires several minutes. This recovery, again, is at first rapid and at the end relatively slow.

Quantitative measurements of these different phases may, as we shall see, be derived from the response-curve itself. In obtaining these, there are two elements to be measured—namely, the extent and the rate of movement. The amplitude or height of the curve gives a measure of the amount of movement. Magnification or reduction of the record results, as we have seen, from two elements of adjustment—namely, the ratio between the horizontal arm of the lever and the length of the recorder, and the ratio between the distance of thread-attachment from the pulvinus and the entire length of the leaf.

In the record given in fig. 12 the length of the vibrating recorder was 10 cm. and the thread-attachment with the leaf was made with the horizontal arm of the lever at a distance of 5 cm. from the fulcrum rod. The magnification of the writing-lever was therefore 2. The total length of the responding leaf was 9 cm. But the thread-attachment to the horizontal lever was made at a point on the petiole 3 cm. from the pulvinus. The responsive movement of that particular point on the petiole was therefore reduced to one-third the movement of the tip of the leaf. Thus we have a reduction to one-third brought about by the selection of the point of attachment on the petiole, and a magnification of two, due to the writing-lever. The record obtained represents in this case the actual movement of the tip of the leaf, reduced to two-thirds.

As regards the time-measurements of the responsive

movement, we have seen that the successive dots in the curve itself give time-intervals. When it is necessary to measure short intervals, say of ·1 second, a resonant vibrator accurately tuned to ten double vibrations per second is employed. The successive dots in the curve then represent intervals of a tenth of a second each.

For the correct determination of the first two phases of the responsive movement in which the time involved is short—namely, the latent period and the apex time—it is necessary to have the recording-plate moving at a rapid rate. But for determining the time-relations during the period of recovery, which is a matter of several minutes, the recording-plate has to be moved at a relatively slow rate.

I give below records (fig. 12) which show these first two elements in a typical manner. The record here, as explained before, is reduced to two-thirds.

Latent Period.—It will be noticed that there is a short interval between the application of stimulus, represented by the vertical line, and the initiation of response. The movement is here seen to begin before an interval of ·1 second is completed. For more accurate determination of the latent period a record must be taken on a faster-moving plate. A detailed description will be found in a later chapter, where it is shown that the average value of the latent period may be taken as about ·1 second. It will also there be observed that though in a given specimen the latent period is constant, it varies slightly in different specimens. It is also appropriately modified according to the physiological changes induced by temperature, fatigue, and the influence of the season.

The Apex Time.—It is seen from the upper of the two curves in fig. 12 that the responsive fall practically attains its maximum near the twentieth dot. This indicates that the value of the apex time in this case is 2 seconds. As regards the rate of this responsive fall, the spacing of the successive dots, each representing an interval of ·1 second, shows in a striking manner how the speed first accelerates and then slows down.

The maximum movement is generally attained about ·5 second after the shock. The actual rate of the maximum responsive fall is here 40 mm. per second. The rate of the responsive fall is modified by various conditions :—

(1) The speed is greater under stronger stimulus. This is well seen in fig. 12, where the lower one was taken under stimulus intensity of 1, and the upper one under stimulus intensity of 4. The gentler slope of the lower curve, and

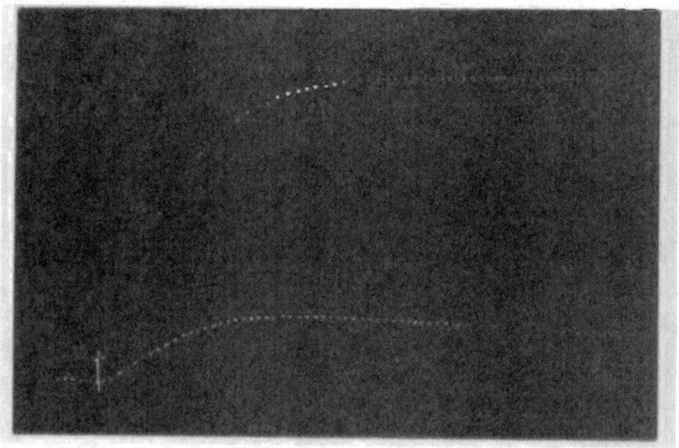

FIG. 12.—Records giving apex-time in the response of *Mimosa*. Lower curve is in response to stimulus 1 and upper to stimulus 4 units.

more abrupt rise of the upper, clearly show the greater speed and vigour of the responsive movement under the stronger stimulus. The curves show moreover that the amount of this responsive movement is greater under stronger stimulation.

(2) In a fatigued condition the rate of the responsive fall under constant stimulus is relatively slow. Thus in a certain experiment, where the maximum rate of fall, when fresh, was 30 mm. per second, the rate was slowed down to 20 mm. per second in consequence of fatigue. In another case the rate when fresh was 50 mm. per

second, which under pronounced fatigue was reduced to 8 mm.

(3) Temperature also modifies the rate of the responsive movement. Thus in a given specimen the maximum rate of fall at the relatively low temperature of 22° C. was 10 mm. per second; it became enhanced to 105 mm. per second at 25° C., and 115 mm. per second at 31° C.

The pulvinus after the attainment of the maximum fall remains more or less persistently contracted for a short time. This is shown by the horizontal portion of the curve in fig. 12.

The Period of Recovery.—As this takes a relatively long time, its record has to be made on a slowly moving plate. The unit of time-measurement must therefore be relatively long. If the successive dots were to be made at the ordinary rate of 10 per second, they would become fused and continuous in the record. For this reason I have devised a contrivance by which the successive dots, in the recovery-portion of the curve, are placed at such intervals as to prevent overcrowding. A convenient interval is either 5 or 10 seconds.

The device for producing periodic dots at intervals, say of 10 seconds, consists of clockwork employed to interrupt the current actuating the vibrating recorder at particular intervals. A light six-rayed wheel is attached to the axis of the seconds-hand, and during the course of a single complete revolution, that is to say, in a minute, the projecting rays press the spring-key six times at intervals of 10 seconds each (fig. 13). It is only during the short interval when the key is pressed that the circuit is completed and the recorder set in vibration to make its dots. Each dot made, it should be remembered, is the result of a succession of strokes inscribed by the vibrating recorder 10 times in the second. But as the movement of the plate is slow, these successive strokes more or less superimposed make but a single large dot. Ten seconds again elapse before the next pressure of the key brings about another large dot, and in that interval the plate has moved a certain distance. There is a second key used for short-circuiting by which the clock interrupter can be put out of action. When this is done, we obtain the usual series of dots at intervals of ·1 second. It will be understood that if the whole response-record were to be made by the vibrator, as actuated periodically at intervals of 10 seconds, we should have a very long gap in the

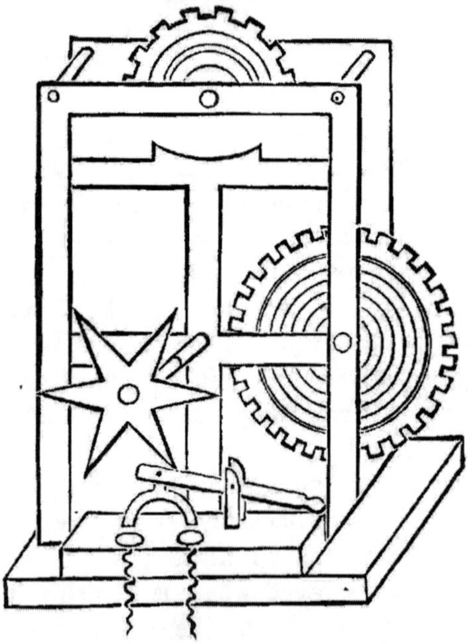

FIG. 13.—Clockwork for the dot-marker; the six-rayed wheel periodically completes electric circuit.

record of the contraction-portion, since the apex is reached in 2 seconds. To obtain then a more or less continuous inscription of the entire curve, the vibrator should at first be allowed to make its normal dots ·1 second apart. This is done by taking care to commence the experiment with the clock-interrupter short-circuited. As soon as the apex point is reached the short-circuit is removed, and the succeeding record, during the recovery of the leaf, consists of dots at intervals of 10 seconds. Fig. 14 gives us a record of a response reduced to two-thirds taken in this way. It will be seen that the recovery practically takes place in 16 minutes. Thus in the present case, while the pulvinus took only 3 seconds to complete the contraction, it required

FIG. 14.—Response of *Mimosa*. Successive dots are at intervals of $\frac{1}{10}$ second in the contractile portion, and 10 seconds in the recovery portion of curve. Vertical marks below indicate intervals of 1 minute.

16 minutes to recover from it. It will be seen, further, that the rate of recovery is quicker at the beginning and slower at the end. The following is a tabular statement of the time-relations of the different phases of response and recovery :—

TABULAR STATEMENT SHOWING TIME-RELATIONS OF RESPONSE AND RECOVERY IN LEAF OF *Mimosa*

Period of contraction	3 seconds
„ recovery	16 minutes
Maximum rate of contractile movement	24 mm. per second
„ „ movement of recovery	.09 „ „
Average rate of contractile movement ..	15 „ „
„ „ movement of recovery ..	.045 „ „

We may now briefly recapitulate the sequence of events in a typical specimen of *Mimosa* subjected, during the summer season, to a moderate stimulus. Response does not commence immediately; there is a latent period of ·1 second. The responsive movement then begins and proceeds for a time with increasing speed, the maximum contraction being attained about 3 seconds after the shock. The pulvinus remains in the contracted position for a short period. After this the recovery is initiated. The rate of recovery at the beginning is relatively rapid, and very slow towards the end. The maximum rate of recovery is ·09 mm. per second, in contrast with the maximum rate of contraction, which is 24 mm. per second. The movement of recovery is thus about three hundred times slower than the movement of contraction. The recovery is completed in about 16 minutes.

A stronger stimulus, generally speaking, requires a longer period for recovery. The influence of season is also a factor to be taken into consideration. Under the physiological depression induced by winter, the responsive process is appropriately modified. The excitability of the tissue becomes depressed. An intensity of stimulus which in summer was effective, becomes in winter ineffective. To evoke response much stronger stimulus has to be employed. The latent period is prolonged and the amplitude of response reduced. And lastly, in winter there is, generally speaking, a great prolongation of the period of recovery. In summer, with vigorous specimens, recovery may be practically complete in as short a time as 8 minutes. But owing to sluggishness induced in winter, on the other hand, the recovery may be prolonged to 25 minutes or more. In a severe winter response may even be abolished altogether.

I have hitherto dealt in some detail with the responsive movement of *Mimosa*. In contrast with this may be cited other examples in which the excitatory reaction may be either more rapid or extremely sluggish.

Response of Biophytum.—As an instance of relatively quick reaction I give (fig. 15) a record of response of leaflet of *Biophytum*. The maximum fall was here attained in the course of a second after the shock. The recovery was completed in the course of only 3 minutes.

Response of Neptunia.—In marked contrast with the quick reaction of *Biophytum*, I may cite the very slow action

FIG. 15.—Record of response of leaflet of *Biophytum*. Vertical marks below record indicate intervals of ·5 minute.

of the primary leaf of *Neptunia oleracea*. In fig. 16 is given a record of its response under an exciting induction-shock of moderate intensity. The clock-interrupter was so adjusted that the successive dots should be at intervals of half a

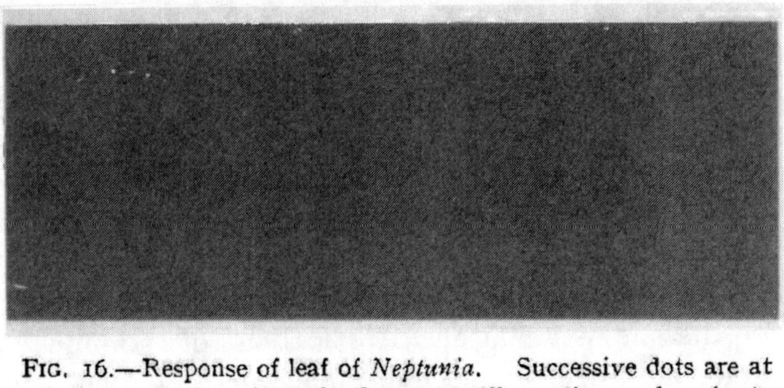

FIG. 16.—Response of leaf of *Neptunia*. Successive dots are at intervals of ·5 minute in the contractile portion, and 1 minute in the recovery portion of curve.

minute during contraction, and at intervals of a minute during recovery. It will be seen that the maximum fall was attained 3 minutes after stimulation, and the recovery was not completed even after 40 minutes, which was the duration of this particular record. The recovery was completed after a further period of 20 minutes, that is to say, the total period of recovery was an hour. The apex time or period of contraction is shortened by the application of a stronger stimulus, but the period of recovery then becomes very much prolonged. The following tabular statement will display the range of variation in the speed of reaction of these sensitive plants:—

TABLE SHOWING APEX-TIMES AND PERIODS OF RECOVERY IN DIFFERENT PLANTS

Specimen	Apex time	Period of recovery
Biophytum sensitivum ..	1 second	3 minutes
Mimosa pudica	3 seconds	16 minutes
Neptunia oleracea	180 seconds	60 minutes

ARBITRARY DISTINCTION OF SENSITIVE AND ORDINARY PLANTS

The arbitrary distinction that is generally drawn between the so-called sensitive and ordinary plants may briefly be referred to here. In the case of *Mimosa*, it is generally supposed that the lower half of the pulvinus is alone sensitive; this however is not an accurate statement. By local application of stimulus it can be shown that the upper half also undergoes a feeble contraction, causing an 'up' movement of the leaf.

The localised stimulus may be applied by means of the electro-thermic stimulator. Application of stimulus of moderate intensity on the upper half of the pulvinus will be found to give rise to an erectile response. Another practical method of local application of stimulus is by means of sunlight. A narrow beam may thus be thrown on the upper half of the pulvinus; this will be found to give rise to erectile or 'up' response. Two such responses are shown in fig. 17, where the stimulus of sunlight was applied for 2 minutes followed by a period of recovery for 13 minutes.

If the stimulus applied on the upper half be strong or long continued, then the excitatory effect is transmitted across the pulvinus to the more excitable lower half. In these circumstances the 'up' is converted to 'down' response, on account of the greater contraction of the lower half of the pulvinus. Thus under any form of diffuse stimulation the resultant response in *Mimosa* is brought about by the differential excitabilities of the upper and the lower halves of the pulvinus. We should also bear in mind that the slight differential contraction-effect in *Mimosa* leaf is very much magnified by the long petiolar

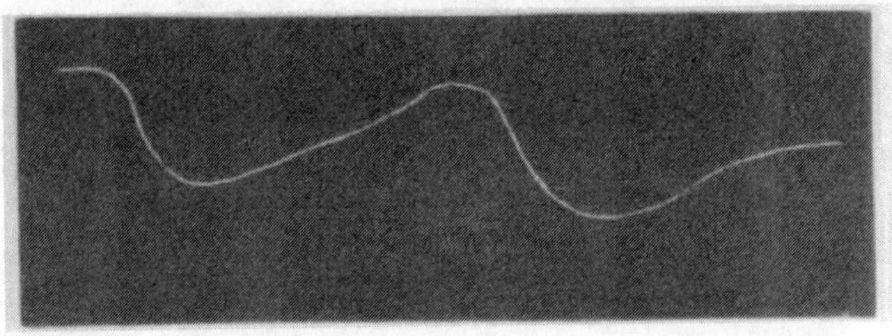

FIG. 17.—'Up' response (represented by down curve) due to local stimulation of upper half of pulvinus of *Mimosa*.

index. There are, again, numerous pulvinar organs whose responsive movements have passed unnoticed. In *Desmodium gyrans* there are two conspicuous pulvini; the primary pulvinus is at the junction of the petiole with the stem; there is a secondary pulvinus at the junction of the petiole with the terminal leaflet. The primary pulvinus appears at first sight to be insensitive. But on attaching the primary petiole of *Desmodium* with the writing-lever, I obtained the series of responses under a very feeble electric shock, as seen in fig. 18. In this particular case the recovery is practically complete in 15 minutes. Other pulvini also exhibit differential contraction under diffuse stimulation. Thus the terminal leaflet of the bean plant (*Vicia Fava*) exhibits responsive down movement, though here recovery is very protracted. But we have seen that the recovery of *Neptunia* is also a very slow process.

The responsive movement of *Mimosa* is due, as has been noted, to the unequal excitabilities of the upper and lower halves of the pulvinus. The excitability of the tissue is again modified by the state of turgor. In *Mimosa* there is induced a periodic variation in the relative turgescence of the two halves of the pulvinus. On account of this the differential excitability, on which the motile response of

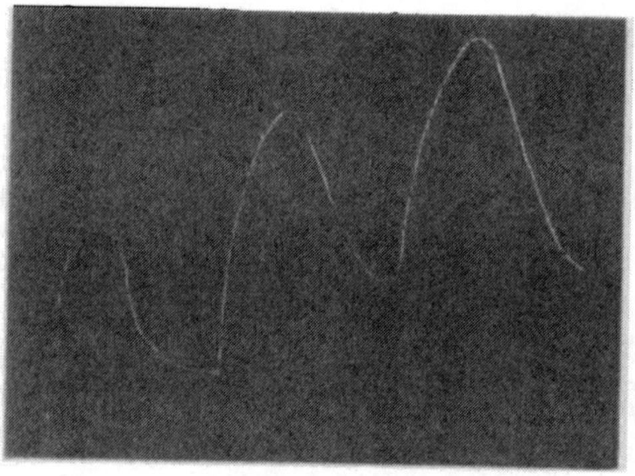

FIG. 18.—Series of responses of leaf of *Desmodium gyrans*, under electrical stimulation.

Mimosa depends, undergoes great variation. The sensitiveness of this plant is in consequence often found to disappear completely at certain hours of the day. I shall, moreover, show in Chapter VII that the leaf of *Mimosa* becomes insensitive when its pulvinus absorbs an excess of water. Thus the mechanical movement of the sensitive plants on which depended the assumption that 'ordinary' plants were insensitive, rests on a basis which is very unreliable.

Responsive movements may, on the other hand, be demonstrated in ordinary plants by the employment of a suitable contrivance. In a radial organ diffuse stimulation induces equal contractions on all sides, which balance each other. Hence lateral movement, dependent on differential contraction, cannot take place. But if we take a hollow tubular organ of some ordinary plant, say the peduncle of daffodil, it is clear that the protected inner side of the tube must be the more excitable. When this is cut in the form of a spiral strip and excited by means of an electric shock, we observe a responsive movement to take place by *curling*, due to the greater contraction of the inside of the strip. This mechanical response is at its maximum at that season which is optimum for the plant. When the plant is killed its response disappears.

It will be seen that the division of plants into sensitive and insensitive is without any justification. Moreover, by adopting the electric mode of investigation, I have shown that every plant and every organ of the plant is sensitive and responds to stimulus by a definite electric variation.

We have hitherto referred but vaguely to the question of the intensity of the induction-shock employed as stimulus to induce response. We have observed that on making and breaking a current in the primary coil, instantaneous currents are induced in the secondary. The intensity of the induction-current employed for giving a shock depends in the first place on the intensity of the primary current; secondly, on the suddenness with which the primary current is made or broken; and lastly, on the relative distance separating the secondary from the primary coil. The intensity of the current can be maintained uniform if we always employ the same battery, say a 4-volt accumulator or storage-cell. As the break of a current is accomplished more quickly than make, the break-shock, as we have seen, is more intense than the make-shock. The plant may, therefore, be excited by a single make-shock, or by a single break-shock or a double make-and-break shock, or by a sequence of make-and-break shocks, of definite duration, according to the particular requirements of the experiment.

The intensity of the shock moreover may, as already

shown, be increased by sliding the secondary nearer and nearer to the primary coil. At a great distance the intensity of the shock is very feeble, whereas in the nearest position it is most intense. If a scale be placed to mark the relative position of the secondary to the primary, we may be assured of obtaining an identical intensity of shock whenever we place the secondary at the same point on the scale; or we can obtain an increasing intensity of stimulus by progressive movement along the scale towards the primary. There is, however, no simple relation between the distance and the intensity—that is to say, equal decrement of distance does not mean equal increment of intensity. All that we are sure of, is that the sliding in of the secondary coil secures an increasing intensity of stimulation. In order to be certain of obtaining quantitative values of intensity, the scale has to be specially calibrated.

Standardisation of Stimulus

In subjecting the plant to the secondary shock, if we begin with feeble intensity of stimulus, by placing the secondary at a great distance, and gradually increase the intensity by sliding the secondary nearer and nearer, we shall obtain that scale-reading at which the stimulus begins to be effective. This particular intensity, the feeblest that is effective, we designate the *minimal stimulus*. As we now proceed to increase the stimulation by pushing the secondary nearer to the primary, we find the amplitude of the response is progressively enhanced, and ultimately we reach an intensity beyond which there is no further increment of amplitude. This intensity we designate the *maximal stimulus*. When the plant is in an exceedingly vigorous condition, the minimal intensity is low and the range between maximal and minimal is narrow. But if the plant be in a less favourable tonic condition, then the minimal stimulus is relatively high and the range between minimal and maximal is wider.

As different induction-coils have different constants, the

value of the intensity of stimulus as obtained from the scale-reading of a particular coil gives us no idea of the absolute intensity. It appeared desirable, nevertheless, in making quantitative experiments, to adopt some unit of stimulus in terms of which other intensities might be expressed. It would be well, moreover, to select this unit in some way not quite arbitrary, so that it might carry a significance more or less universal. The unit intensity of exciting shock which I have adopted for these reasons is that which barely induces in ourselves a perceptible sensation. The observer dips two fingers, one of each hand, into two troughs of saline solution, which are in series with the experimental *Mimosa* and the secondary coil. The plant tissue is interposed so as to ensure an identical current to pass through the experimental individual and the plant. The resistance offered by the plant tissue is very great; in the case of *Mimosa* under the usual mode of connection, it is about half a million ohms. At the beginning the secondary is placed at a great distance from the primary. The vibrating interrupter of the primary is next started and the secondary gradually pushed in, till at a certain scale-reading the observer, who is kept in ignorance of the position of the secondary, just begins to perceive the shock. This process is repeated several times in the case of the individual observer, and the mean of various consecutive readings, which ought not to differ from each other to any extent, is taken as the unit for that particular individual. The same observation is repeated with some ten different individuals, and the mean of these ten readings is finally adopted as that reading of the unit intensity which is to serve as the standard.

Though this reading cannot be regarded as absolute and invariable, yet, in the particular circumstances of the case, it is fairly definite and on the whole satisfactory. It gives us a general idea, moreover, of that intensity which will be effective in stimulating the plant, in terms of the minimal stimulus capable of evoking sensation in man. Having thus obtained the scale-reading corresponding to this unit,

E

we calibrate other positions of the scale in terms of this unit. In this manner the scale is marked so as to indicate intensities of ·1, ·5, 1, 2, 3, 4, 5, and so on. The calibration is carried out by means of a ballistic galvanometer. In subsequent chapters we shall employ these practical units, which will thus have a definite significance.

Having shared the prevailing belief that the sensitiveness of the plant was very feeble compared with that of the animal, I was considerably surprised to find that the intensity of induction-shock which is barely sufficient to induce sensation in man is quite enough to cause excitatory fall in a *Mimosa* of moderate sensitiveness. Indeed, I found that in the case of a highly excitable specimen an intensity only one-tenth of this was sufficient to excite it. In other words, under this particular test *Mimosa* may prove ten times as sensitive as a human subject! Later on I shall give details of measurements which will show that, as far as electric mode of stimulation is concerned, the plant is in no way inferior to the animal in sensitiveness.

SUMMARY

The extent of responsive fall in *Mimosa* increases with increasing intensity of stimulus. The rate of movement is also greater under stronger stimulus.

The rate of responsive movement becomes slower under fatigue. In a given case the normal maximum rate of movement of 50 mm. per second was reduced to 8 mm. under fatigue.

Temperature enhances the rate of movement. A rate of 10 mm. per second at a temperature of 22° C. was found enhanced to 105 mm. per second when the temperature was raised to 28° C.

In a typical case of *Mimosa,* in summer, the latent period was found to be one-tenth of a second. The maximum contraction was attained in 3 seconds and the recovery completed in 15 minutes. The rate of recovery was relatively rapid at the beginning and very slow towards the end. The maximum rate of recovery was ·09 mm. per second in contrast with the maximum rate of contraction of 24 mm. per second. The movement of recovery was about three hundred times slower than the movement of excitatory contraction.

A stronger stimulus, generally speaking, requires a longer period for recovery.

Under the physiological depression induced by winter the responsive reactions are modified. The latent period is prolonged and amplitude of response reduced. The period of recovery may also become protracted.

Different plants exhibit different characteristics of response. *Biophytum sensitivum* may be taken as a type of quickly reacting plant, while *Neptunia oleracea* is very sluggish in its reactions. In *Biophytum* the apex time is reached in a second and the recovery accomplished in 3 minutes. In *Neptunia* the apex time is reached in 180 seconds, and recovery completed in 60 minutes.

Mechanical response of *Mimosa* is due to differential contraction of the upper and lower halves of pulvinus.

Erectile responses of *Mimosa* may be obtained by local stimulation of the upper half of pulvinus.

Distinction of plants into sensitive and ordinary is arbitrary. Under suitable conditions, ordinary plants, so-called, may be made to exhibit motile response.

By means of electric response it may be shown that every plant, and every organ of the plant, is sensitive and responds to stimulation by a definite electric change.

The sensitiveness of *Mimosa* to electrical stimulus is high and may even exceed that of a human subject.

CHAPTER V

THE ADDITIVE EFFECT; INFLUENCE OF LOAD, TEMPERATURE, AND INTENSITY OF STIMULUS

Greater excitatory efficiency of the break-shock—Additive effect of stimulus—Quantitative relation of additive effect—Effect of load—Measurement of work under different loads—Rate of work—Thermal chamber—Effect of temperature—Effect of increasing intensity of stimulus on response.

IN exciting *Mimosa* by means of induction-currents we may employ either the make- or break-shock. It has already been stated that the break-shock is more efficient than the make-shock. That is to say, as we gradually push in the secondary nearer the primary, excitation is effected earlier with the break than with the make. I will now proceed to demonstrate this fact by experiments.

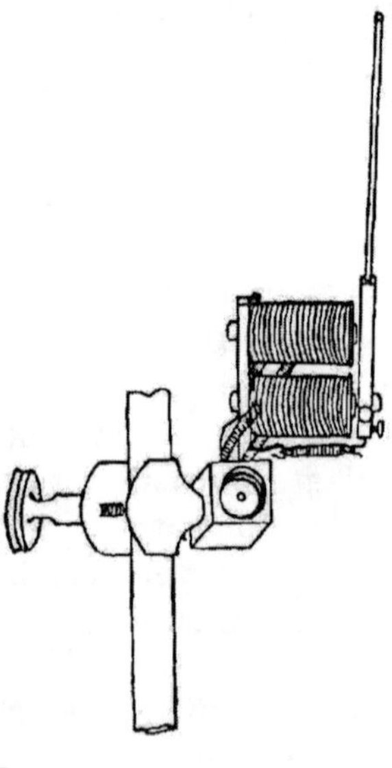

FIG. 19.—The electric signal.

For obtaining the record I employed a writer which had a vibration-frequency of 20 times per second. The make and break of the primary current was effected by a metronome. In the primary circuit an electrical signal (fig. 19) was also included, which marked at the base of the figure the moments when the current was made and broken. When the current is made, an up-line is described by the writer attached to the signal. So long as the current is flowing, the writer remains in the up-position and draws a horizontal line (fig. 20). At the time of make it will be noticed that, owing to inertia, the writer was momentarily jerked somewhat above the level of this up-position. This jerked line, therefore, always marks the moment of make, and the horizontal line at the higher level the continuation of the current. When the current is broken, the writer falls suddenly to its original level. Thus a jerked up-line indicates the moment of the application of the make-shock,

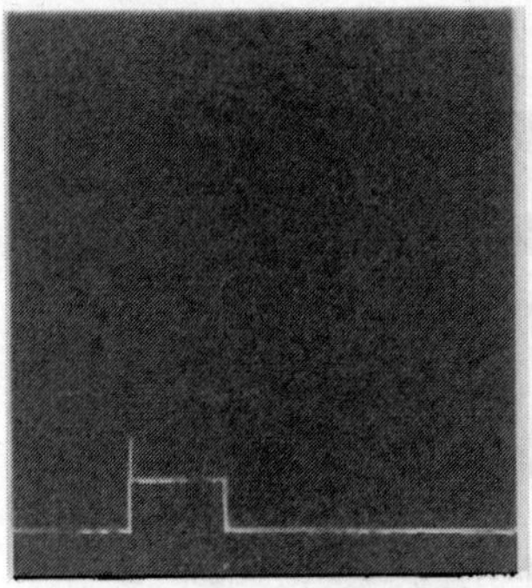

FIGS. 20, 21.—Records showing greater efficiency of break-shock; frequency of vibrating recorder is 20. Signal below shows by up movement 'make' and by down movement 'break.'

and the down-line the application of the break-shock. In the two accompanying figures are given records of the effects of make- and break-shocks.

GREATER EFFECTIVENESS OF BREAK-SHOCK

In the record (fig. 20) the secondary coil was placed at the reading of ·75 unit. It will be noticed that at 'make' there was no response. But there was response at 'break,' which took place ·1 second later, the delay being due to the latent period. In the next experiment, with the same plant, the coil was pushed into the reading of 1. It will be seen (fig. 21) that excitation was here effective at 'make,' a similar delay of ·1 second being again due, as in the previous case, to the latent period. Thus we see that while the stimulus of the feebler intensity of ·75 was effective at 'break,' it took the stronger stimulus of 1 to induce response at 'make.'

ADDITIVE EFFECT OF STIMULUS

In the responsive tissue of the animal a single stimulus, by itself ineffective, is found to become effective on repetition. In order to test whether this holds good in the case of the plant also, I carried out the experiments which I

FIG. 22.—Stimulus of intensity ·5 became effective on being repeated four times.

shall now describe. With a given specimen I found that a single make-and-break shock of intensity ·75 was ineffective in inducing excitation. I then adjusted the secondary for intensity of ·5, and made a reed-interrupter interposed in the primary circuit give a series of make-and-break shocks till the leaf responded by a fall. The interrupting reed was adjusted to vibrate five times per second and the number of interruptions is recorded below in the usual manner. It will be seen in the record given in fig. 22 that the make-and-break stimulus, which singly was ineffective, here became effective on being repeated four times.

Desiring next to observe the effect of still further reducing the intensity of stimulus with the same specimen, I adjusted the secondary for an intensity of ·1. It must be remembered that this is the intensity of tetanisation, which is only one-tenth of what is perceptible to the human subject. Looking at fig. 23 it will be seen that even this very feeble stimulus became effective on being repeated 20 times.

In carrying out this experiment I had expected in a general way that a feeble stimulus, to be effective, must be repeated a greater number of times. But I was not prepared for so strictly quantitative a result as came out in these two records. If the summated effect is to prove strictly additive, then effective excitation must be equal to the individual intensity multiplied by the number of repetitions. From the record in fig. 22 the effective excitation

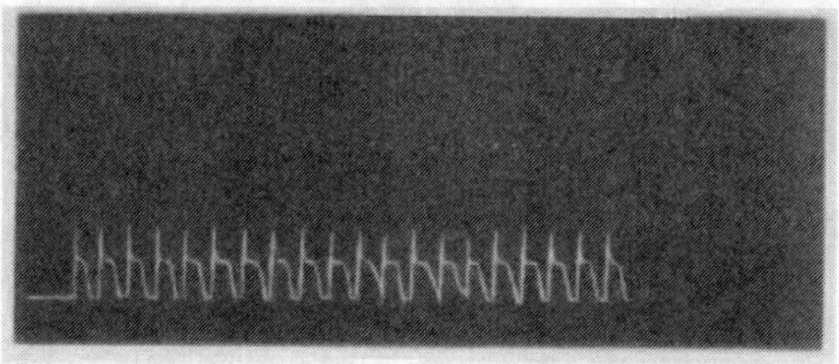

FIG. 23.—Stimulus of intensity ·1 became effective on being repeated twenty times.

was seen to be $\cdot 5 \times 4 = 2$. From the second record with the same specimen, in fig. 23, it is seen to be $\cdot 1 \times 20 = 2$. In other words, for effective excitation the number of additive stimuli varies inversely as the intensity of each. That this is true, within certain limits, is borne out by another set of results obtained from a different specimen, which was found to be somewhat more excitable than the former.

In order to vary the condition of the experiment I adjusted the reed-interrupter to vibrate twice in a second. There was thus an addition here of the effects of single make-and-break shocks at intervals of half a second, instead of one-fifth of a second as in the last case. In fig. 24 is seen the record of the additive effect, the intensity of stimulus being ·5. We find here that the stimulus became effective on being repeated twice.

The experiment was again repeated with the same

specimen, but using the reduced stimulus-intensity of ·2. The result given in fig. 25 shows that the stimulus had to be repeated five times to become effective. We see once more in this experiment that the additive effect is strictly quantitative, and that the effective stimulation is constant under varying intensity of stimulus, being equal to the individual intensity multiplied by the number of repetition. In the

FIG. 24.—Stimulus of intensity ·5 became effective after two repetitions.

FIG. 25.—Stimulus of intensity ·2 became effective after five repetitions.

two cases here given we have a strict reaffirmation of this quantitative relation—namely, $\cdot 5 \times 2 = \cdot 2 \times 5 =$ constant.

INFLUENCE OF LOAD

In the response of muscle it is found that the muscle-curve is modified by the effect of the load which it has to raise during contraction. With an increasing load the height of response undergoes a progressive diminution, but the period of recovery is at the same time correspondingly shortened. In the contractile response of *Mimosa* a similar phenomenon is observed. In carrying out this experiment a load was placed on the arm of the horizontal lever opposite to that of the leaf-attachment, and at an equal distance from the fulcrum. The leaf, during its contractile movement, has to lift this weight. In the first experiment of the series a load of 100 milligrammes was employed. In the second this was increased to 500 mgrms., and in the last it was made 2000 mgrms. Successive records were made, for purposes of comparison, on the same part of the plate. The vertical lines under the diagram (fig. 26) are time-marks, indicating intervals of one minute. It will be seen that the height of the record, with a load of 100 mgrms. is the greatest of the three, being 41 mm. The recovery was completed in this case after the expiration

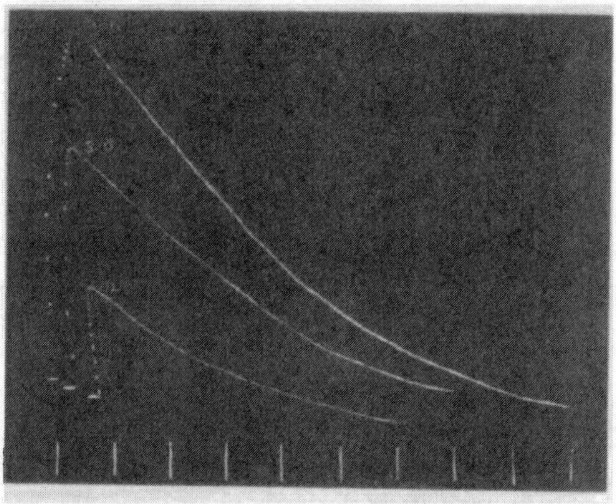

FIG. 26.—Effect of load; the three records show responses under varying loads of 100, 500, and 2000 mgrms.

of 9 minutes. The height of the second, under a load of 500 mgrms. is 28 mm., but recovery is nearly complete in 6.5 minutes. And, finally, with the load of 2000 mgrms. the height is the least, being 13 mm., but the recovery is seen to be completed in 5 minutes. Thus the effect of load in the contractile response of the plant is shown to be strictly parallel with its influence on the contractile response of animal muscle.

WORK PERFORMED BY THE PLANT

The motile tissue of the plant, like that of the animal, is capable of doing work during excitatory movement. The influence of load on the height and period of response has been shown to be similar in the two cases. We may next study the effect of load on the work performed. Work is measured by the product of the weight raised and the height of the lift. In the response of muscle, if the increasing loads are represented by W_1, W_2, W_3, and the corresponding heights of response by h_1, h_2, h_3, then it is found that up to a certain limit $W_1h_1 < W_2h_2 < W_3h_3$; in other words, the work performed is increased under enhanced load and increasing tension.

Turning to the effect of increasing load on the response of *Mimosa*, we find that with a load of 100 mgrms. the height of response is 41; the value of W_1h_1 is therefore 4100; under a load of 500 mgrms., $W_2h_2 = 500 \times 28 = 14{,}000$; and, lastly, with a load of 2000 mgrms., $W_3h_3 = 2000 \times 13 = 26{,}000$. It is thus seen that as in the contractile response of animal, so also in that of the plant, greater amount of work is performed under increased load and higher tension.

We may now try to obtain some idea of the absolute amount of work performed and the rate of work. We shall take the case where the plant had to lift a weight of 2000 mgrms. The following data are available from fig. 26. The weight is seen to be lifted through 12 mm. in the course of ten successive dots, each representing ·1 second. The magnification of the lever was three times; the absolute lift is therefore 4 mm. The load to be lifted was 2000 mgrms.; but the weight of the leaf was 130 mgrms. and this helped the fall. The actual work performed is therefore $(2000 - 130) \times 4$ millimetre milligrams. This was accomplished in the course of a second. Hence the absolute rate of work was 7480 mm. mgrms. per second.

Effect of Temperature

Our next inquiry is into the effect of temperature on the response of the plant. For this we have to subject the plant to different temperatures—some low, some high—and to find means of maintaining it constant at any definite temperature required. For this purpose a plant-chamber, enclosing the plant and fulfilling these conditions, had to be devised. It will be noticed that these investigations involve two opposite sets of requirements—namely, in the one case a definite lowering of the temperature of the chamber below, and in the other a definite raising of it above, the temperature of the environment.

The plant-chamber consists of a base-board with a rectangular cover. This cover is made of a light wooden framework, the sides being closed with sheets of mica. The advantage of mica is its lightness, unbreakableness, non-conductivity, and transparency. Transparency is necessary because in darkness the sensitiveness of a plant undergoes variation. The base-board consists of two halves, with a small circular opening in the middle. When these two halves of the base-board are slipped over the top of the flower-pot they form one piece, fixed together by means of suitable clasps. The base-board rests on the flower-pot and the main stem of the plant passes through the circular opening. The base-board thus forms the floor of the thermal chamber. There are grooves cut in the base-board for the reception of the wooden framework. The plant is thus enclosed except on the top. After making the necessary thread-connections of the lever with the responding leaf, the top is closed by means of two sliding-pieces of mica, with slits for the passage of the thread. There are two side-tubes, one near the top and the other near the base, for the passage in and out of a stream of cold air, when the temperature of the chamber is to be reduced. When the temperature is to be raised, an electrical heating arrangement is employed.

The requirements of cooling are, first, a weighted air-bag, provided with a stop-cock; and second, a coiled copper-pipe placed in an ice-box. By means of indiarubber tubing, connections are made, first, between the stop-cock of the air-bag and one end of the copper pipe; and, second, between the other end of the copper pipe and the upper tube of the thermal chamber. Thus by more or less opening the stop-cock of the air-bag a stream of cooled air is made to circulate through the plant-chamber, at varying rates. A steady low temperature may thus be attained by adjusting the inflow of cooled air, the degree of cooling being dependent on the rate of flow. A thermometer placed inside the chamber indicates the temperature attained.

In order to raise the temperature of the plant-chamber an electrical device is employed. Inside the rectangular frame there is a coil of wire of German silver, the ends of the wire being led outside to two binding-screws. An electrical current from an outside battery is led through this wire, a variable resistance being also interposed in the circuit. The heat generated inside the chamber can be increased or decreased by changing the intensity of the current; this is accomplished by varying the adjustable resistance. In this manner it is quite easy to raise the temperature inside the plant-chamber to any degree that is desired, and to maintain it constant as long as necessary.

FIG. 27.—Effect of temperature; amplitude of response seen to be higher, and period of recovery shorter, with higher temperature.

The temperature of the room, at the time of the experiment I am about to describe, was 27° C. I desired to take three records, differing from each other by intervals of 5° C. For this purpose I reduced the temperature of the plant-chamber to 22° C. and took the first record of the series. Next, by stopping the inflow of the cooled air and opening one of the side windows, I restored the temperature of the chamber to 27° C., and after allowing a suitable interval took the second record. Lastly, by means of the electrical heating device described, the temperature of the chamber was raised to 32° C. and the third record of the series taken. It should be mentioned that in all these cases the stimulus employed was of constant intensity—namely, 2.

In fig. 27 are shown the responsive effects of an identical stimulus at these three different temperatures. At 22° C. it is seen that the height of response is small and the recovery extremely prolonged. At 27° C. we find the amplitude of response enhanced and the rate of recovery increased. At

FIG. 28.—Response taken at three different temperatures, the lowest below and the highest above, on a faster-moving plate; amplitude of response larger and steepness of curve greater, at higher temperature.

32° C. the height of response is still more enhanced and the rate of recovery, as seen in the steepness curve, still further increased. In fig. 28 is given another set of records taken on a faster-moving plate, exhibiting the effect of temperature on the amplitude of response. It will be shown in a succeeding chapter that the latent period also is affected, being progressively decreased with rising temperature.

Influence of Stimulus-intensity on the Response

It is usually supposed that in *Mimosa* every effective stimulus causes the maximum response. That this is not the case comes out very clearly in careful records taken with gradually increasing stimuli. We have already seen, in fig. 12, the marked heightening of the response under an increased intensity of stimulus. In muscle, in the narrow range between minimal and maximal stimulation, there is increasing amplitude of response with increasing stimuli. But this soon attains a limit beyond which there is no further increase of responsive contraction, whatever be the stimulus-intensity employed.

In order to demonstrate a similar progressive increase in the response of *Mimosa*, I first determined the minimal

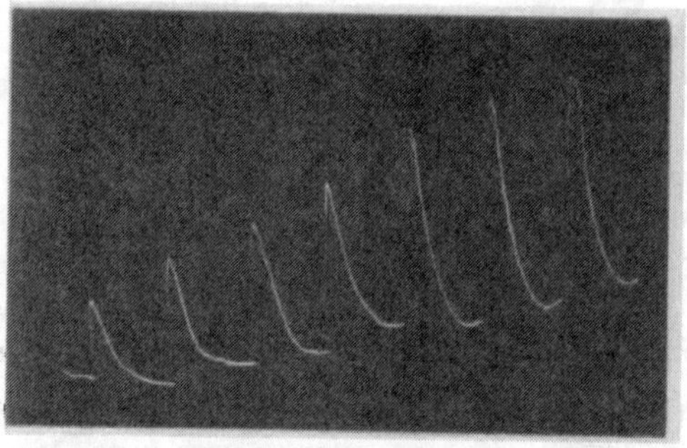

FIG. 29.—Increasing response under increasing intensity of stimulation.

stimulus that was barely effective in inducing a feeble response. Starting from the particular position of the secondary which gave this minimal intensity of stimulus, I very gradually increased the intensity, by moving the secondary only 5 mm. at a time nearer to the primary. At each step I took a corresponding record.

In fig. 29 a series of seven such records is shown, the successive responses being taken, as previously mentioned, under slightly increasing stimuli and at intervals of 15 minutes. It will be seen how the height of response is progressively increased. This increase is at first marked, but towards the end we note that a limit is being approached, the difference between numbers 6 and 7 of the series being very slight. After the seventh, it was found that the responses did not undergo any further increase.

The range within which the increasing effect is seen is relatively extended in the case of plants in a somewhat sub-tonic condition. But when the specimen is highly excitable the range of variation is proportionately restricted.

SUMMARY

The break-shock is more effective in inducing excitation than the make-shock.

Stimulus, singly ineffective, becomes effective on repetition. The effective stimulation is equal to the individual intensity of stimulus multiplied by the number of repetitions.

The effect of load on the response of *Mimosa* is similar to that on the contractile response of muscle. With increasing load the height of response undergoes a progressive diminution with shortening of period of recovery.

Within limits, the amount of work performed by a muscle increases with the load. The same is true of work performed by the pulvinus of *Mimosa*.

In a given case the rate of work performed by the pulvinus of *Mimosa* was 7480 mm. mgrms. per second.

The effect of rising temperature on response is to enhance the amplitude and to shorten the period of recovery.

In *Mimosa*, increasing intensity of stimulus induces increasing amplitude of response. This, however, soon reaches a limit.

CHAPTER VI

VARIOUS TYPES OF RESPONSE

Necessity of uniform stimulation—The Periodic Starter—The Automatic Exciter—Electrolytic contact-maker—The complete Response-recorder—The factor of tonicity—Uniform responses—Fatigue under shortened period of rest—Growing fatigue—Alternating fatigue—Staircase response—Explanation of erection of leaf under continuous stimulation—Fatigue-relaxation in plant and animal—Response under single stimulus and under tetanisation.

Some of the effects brought about by varying external conditions on the excitability of the plant have now been noted. Certain other variations may, however, be induced in the excitability, in consequence of the after-effect of the stimulus itself, even when the external conditions are maintained constant. We may trace these induced internal changes in the modification of the response-records.

It is clear that we can only be assured of the occurrence of such internal changes from the observed variation of response-record if we have been able in the first place to keep the plant under unvarying external conditions. This, taking certain special precautions as regards light, temperature, and so forth, presents no difficulty. But, in the second place, we have to be specially careful that the testing-stimulus itself shall be absolutely constant in successive experiments. The problem then resolves itself into the successful devising of some arrangement by which records may be taken automatically at definite pre-determined intervals of time. The stimulus of unvarying intensity must also be made to act automatically upon the specimen. Under these conditions any variation that may be observed in the record will be due to changes of excitability induced as an after-effect of the stimulus itself.

Maintaining the stimulus-intensity absolutely constant is not so easy to secure with a single make- or break-shock, since the intensity of such a shock is liable to variation, according to the degree of suddenness with which it is effected; but the total additive value of a group of such shocks may be expected to be fairly constant. For this reason, therefore, tetanising shocks caused by a vibrating interrupter would be preferable, provided the duration of these shocks, depending on the duration of closure of current in the primary circuit, be maintained in successive experiments rigorously equal. Such constancy cannot be arrived at if the closure of the circuit be caried out by hand, or even by metronome. Some special mechanical device must therefore be adopted for this purpose.

It is further necessary, in order to maintain constancy of conditions, that identical periods of recovery should be allowed in successive records. For this the stimulus must be applied at accurate and pre-determined intervals of time. The ideal condition, then, for the final elimination of all uncertainties due to the personal factor, is that the plant attached to the recording apparatus should be automatically excited by a stimulus absolutely constant, make its own responsive records, go through its own period of recovery, and embark on the same cycle over again without assistance at any point on the part of the observer.

These demands have been fully met in the devices and adjustments now to be described, consisting as they do of two chief elements—namely, the Periodic Starter and the Automatic Exciter. By the former the time-interval between successive stimulations is regulated; by the latter the stimulation itself, of a definite duration, is applied.

The Periodic Starter

In the case of *Mimosa* recovery from excitation is practically completed in a period of about 10 to 20 minutes, according to the season and the condition of the plant. In practice, therefore, we require arrangements by which successive stimulations can be automatically effected at these intervals. As we require a slowly moving plate for the purpose of these records, the plate-carrier is let down by a thread which is wound round a wheel attached to the minute-hand axis of the driving-clock. To the same axis is also screwed one or other of the three separate discs, bearing equidistant projecting rods, either 3, 4, or 6 in

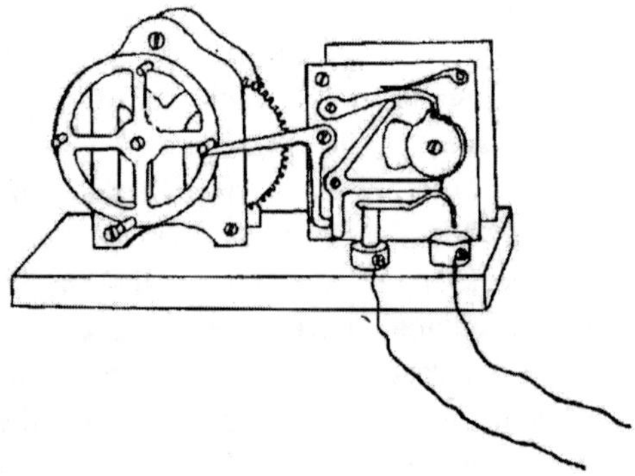

FIG. 30.—Periodic Starter and Automatic Exciter.

number. During one complete revolution, which takes an hour, these rods will press and release a spring at intervals of 20, 15, or 10 minutes, as the case may be (fig. 30).

THE AUTOMATIC EXCITER

If the primary circuit of the induction coil provided with a spring-interrupter be closed for a definite period of time, say ·1 second, then the number of interruptions, with consequent induction-shocks, will also be definite. What is wanted is some contrivance for release, through which the Periodic Starter can close the main circuit for a definite length of time, say ·1 second. It might at first sight appear that this could be secured by an electrical contact made by the revolving radial-rods already referred to, but the period of such contact would be impracticably long—more than 15 seconds.

So continuous a tetanisation would undoubtedly fatigue or even injure the tissue. The contact made by a seconds-hand, again, though sufficiently brief, would have the serious defect that its movements were jerky and would therefore make the duration of contact unequal.

I succeeded in overcoming these difficulties by using a released revolving disc, which could be made to complete an electrical circuit for any definite short period that was required. For this I employed a phonograph motor, an axis of which, carrying a disc, could be adjusted to revolve once in a second.

This disc is usually held stationary by a lever-clutch, and can be started only by the pressure on it of the revolving rod of the Periodic Starter. It is re-arrested after one complete revolution and is not again released till the next rod comes into position, after an interval of 10, 15, or 20 minutes, as the case may be. There is also attached to the disc a sector whose arc is one-tenth the circumference of the circle. This sector, during the revolution, will press against a closing-key, the period of closure being then ·1 second. By increasing or diminishing this arc the time of closure, and with it the duration of the tetanising shock, can be correspondingly changed. It is necessary that the sector should, at the moment of the release, be at the greatest possible distance from the closing-key. By the time it reaches this key it will have acquired a constant and definite velocity. Thus the periods of closure, and consequent duration of the exciting-shock, will be identical in successive experiments.

There is another possible source of variation which must be guarded against. The electrodes of the secondary coil are connected with the plant by means of moistened threads. These threads, in long-continued experiments, may become more or less dried up, the electrical resistance being thus increased in an unknown manner. The intensity of the exciting current may, under these conditions, undergo a change. This difficulty has been overcome by a contrivance for keeping the thread uniformly moist. This will be understood from the diagram (fig. 31) of the electrolytic contact-maker: Two small cells are made of cork; the upper cell is filled with very dilute saline solution, a little of which also lies in the bottom of the lower cell. A bent piece of silver-wire coated with a deposit of chloride, and fixed to the horizontal metallic-rod, pricks through two cells and dips into the solutions above and below. The moistened thread coming through a hole near the bottom of the upper chamber makes one loop round that portion of the plant specimen where electrical connection is desired, turns back into the lower cell (which it enters through the open aperture), and dips into the saline solution. It will be seen that apart from capillary action, owing to the upper chamber being at a higher level, the thread will be kept constantly moist by the slow streaming down of the solution. The current from the coil, again, will have two entries by means of the doubled thread, the resistance being thus halved. The second electrode of the coil is connected with the other contact-point on the plant in a similar manner. The resistance offered by the plant tissue is relatively high, being of the order of a million ohms. The resistance of the electrolytic contacts, on the other hand, need be no higher than a few thousand ohms. By thus making the resistance of the moist contact relatively small, the total resistance of the circuit remains practically the same, especially since we guard against any variation that might

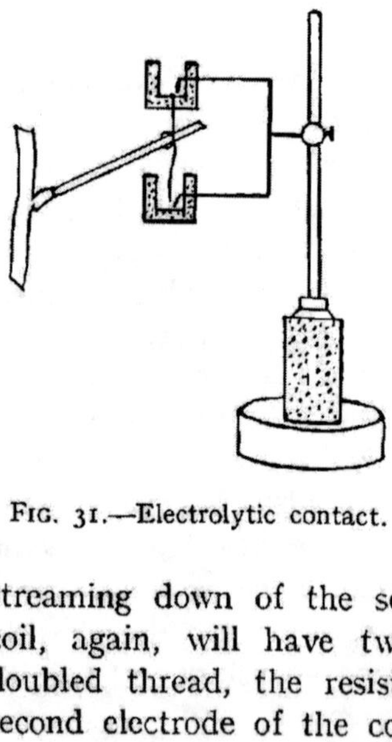

FIG. 31.—Electrolytic contact.

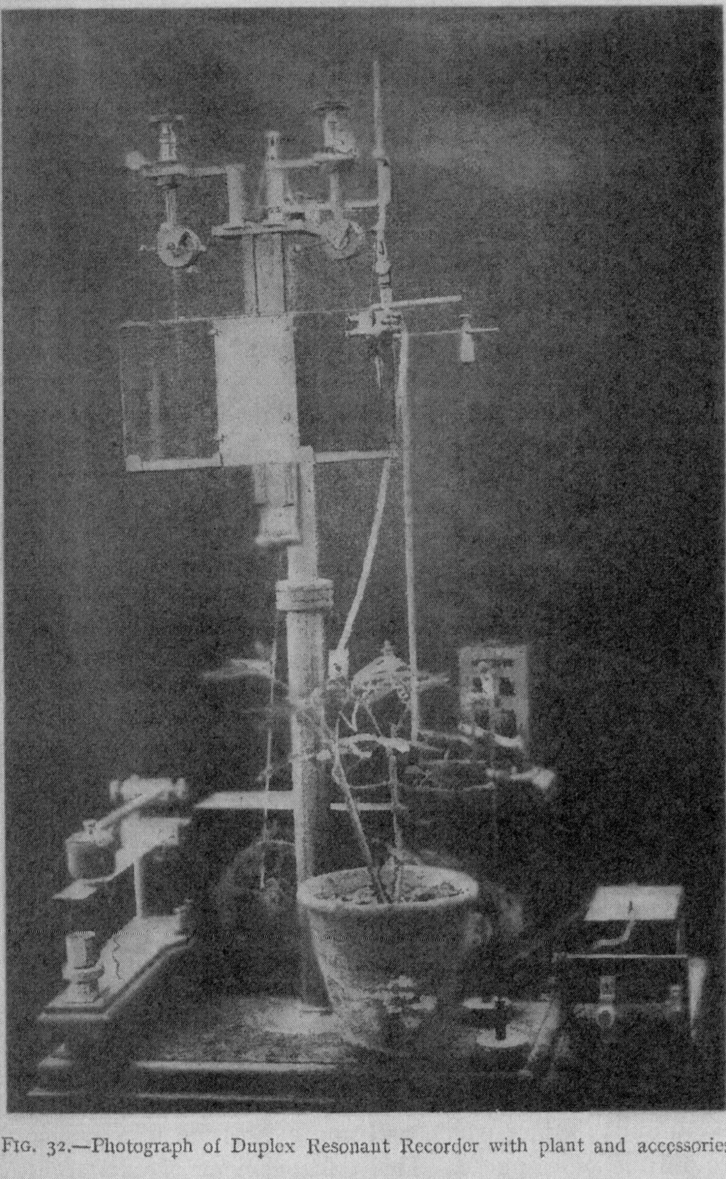

FIG. 32.—Photograph of Duplex Resonant Recorder with plant and accessories.

be induced in the moist thread by drying. The horizontal rod holding the cork of the contact-maker is soldered to a tube which can move up and down a vertical rod. These adjustments enable the point of electrolytic contact to be brought to a level with the point of connection on the specimen. The rod is insulated on ebonite.

All the practical difficulties having thus been eliminated, I shall now proceed to show the various records obtained under this mode of periodic stimulation of uniform intensity. In fig. 32 is given a photograph of the apparatus with its accessories. The recorder is of duplex type, for taking two sets of records at the same time.

Before entering upon the detailed consideration of the results of these experiments, I may say that at the beginning of this investigation my attention was roused by the apparently capricious variations in the responses obtained under conditions which were rigidly uniform.

A long-continued investigation through the different seasons of the year has given me the clue to what at first appeared to be so anomalous. The outward response, it is obvious, is dependent on two factors—the intensity of the impinging stimulus and the capacity for reply possessed by the plant itself. It is easy to see that the second of these factors must be dependent on the vigour of the plant, or in other words, on its tonic condition, which in its turn is modified by the environmental condition. Thus in unfavourable circumstances the plant may fall into an atonic or sluggish condition. The absorption of energy from without, by whatever form of stimulation, will improve the tonic condition of the plant, with consequent enhancement of excitability.

Taking a plant in a subtonic condition, then, we may expect that any application of stimulus will increase its excitability, a fact which will find expression in a growing amplitude of response. This enhancement of excitability will reach a limit at which the plant will be in an optimum condition. After reaching this climax there may be a reversal, with decline of excitability, a state of things which we associate with fatigue.

It must be remembered that in Nature, according to the conditions of its environment, a plant may be found in any of the three states. One specimen may be found in the pre-optimum or subtonic condition; another may be near the optimum condition, and this we shall designate as the normal; a third may be found in the post-optimum condition predisposed to fatigue. The first and third of these conditions may be distinguished from each other by means of testing blows or stimuli. If the plant be in the former condition, these will evoke responses of increasing amplitude; in the latter, they will show a decline.

These three conditions modify not merely the amplitude of response but also exhibit themselves appropriately in other aspects of protoplasmic excitation. These will be seen in the chapters on the *Latent Period*, and on the *Transmission of Excitation*.

Uniform Responses

When—selecting a plant which is neither subtonic nor yet at its optimum—we take a series of responses under uniform stimulation of moderate intensity, allowing sufficient intervals for complete recovery, we obtain uniform responses. This may be accepted as the characteristic effect of a plant in the normal condition.

In fig. 33 is seen a series of such responses taken at intervals of 15 minutes. The ascending portion of each response is here seen to be dotted. This is because of the rapidity of the movement of fall. The successive dots caused by the recorder vibrating ten times per second are widely spaced. In the recovery or down part of the curve, however, as that process is slow, the dots become fused and make a thick continuous line. In the record of the responsive fall, variations of rate of movement may be noticed. At first the speed increases, then very gradually slows down, and the leaf becomes for a time stationary at the apex. These varying rates of fall are seen in the growing and then in the diminishing intervals between the dots.

Fatigue

If instead of giving the full period of rest necessary for complete protoplasmic recovery, the period of rest be

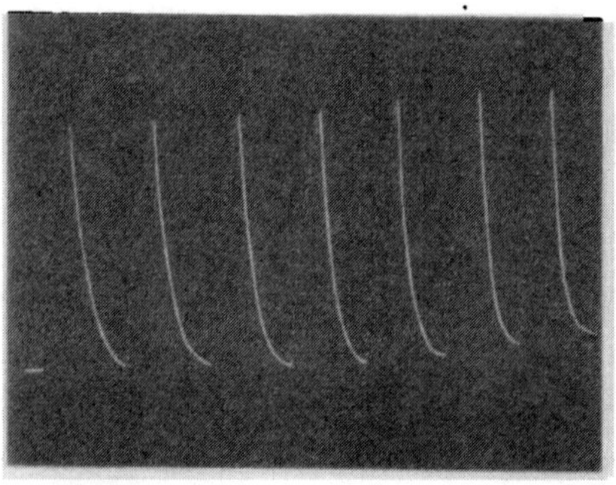

Fig. 33.—Uniform responses of *Mimosa*; stimuli applied at intervals of 15 minutes.

shortened, we obtain a diminution in the height of response indicative of fatigue. This is well seen in fig. 34. The first three uniform responses here—taken, as it is unnecessary to repeat, under uniform stimulation—were recorded at intervals of 15 minutes each. The intervals between successive stimulations were now shortened to 10 minutes, which at once results in a fatigue-diminution of the height of responses. The second three responses appear crowded together, owing to the shortening of the time allowed for record. The time of recovery, after the third of these responses, was again restored to its first value of 15 minutes, and we see at once the reversion of the response to its original height. A similar exhibition of fatigue is also seen in muscle-records, in the same circumstances of diminished interval of rest.

Under certain conditions we obtain an exhibition of continuously growing fatigue. We have seen that when the plant is intensely excited, it takes a longer time for complete protoplasmic recovery. The specimen whose responses are given in fig. 32 happened to be in an optimum condition. A maximum excitation was here induced, even under a moderate stimulus. The normal interval of 15 minutes, which was found in the previous case to be sufficient for complete protoplasmic recovery, here proved to be insufficient. Hence we have the exhibition of a growing fatigue seen in the diminishing heights of successive responses.

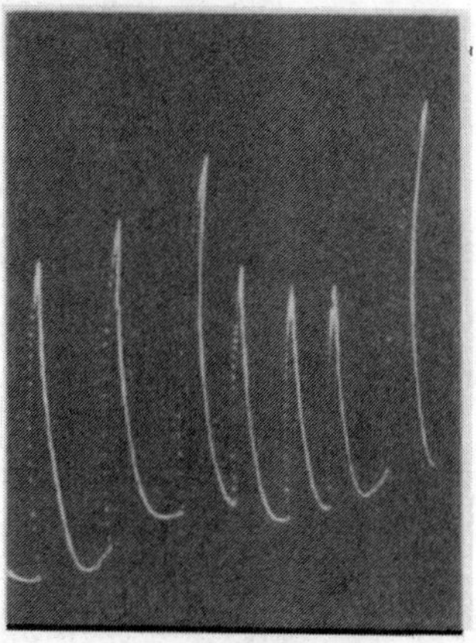

FIG. 34.—Fatigue under shortened period of rest. First three uniform responses obtained at intervals of 15 minutes. The second three, under shortened period of rest of 10 minutes, exhibit fatigue. On returning to interval of 15 minutes, the last record shows enhancement.

Another very curious type of response sometimes met with, is that of alternating fatigue. Here, while the first response is very large, the second is correspondingly small, and this alternating sequence is observed for a longer or shorter time (fig. 36). After several such alternations, however, the responses tended to become uniform. An explanation of this interesting variation may be gathered from careful observation of the record. The freshness of the specimen and its high excitability account for the great amplitude of the first response. An intense excitation requires, as we have seen, a correspondingly longer time than does a feeble one for complete recovery. Hence in the present case the second stimulation is seen to have impinged on the organ before complete protoplasmic recovery has had time to take place, during the usual resting-interval of 15 minutes. The consequence of this is the diminished

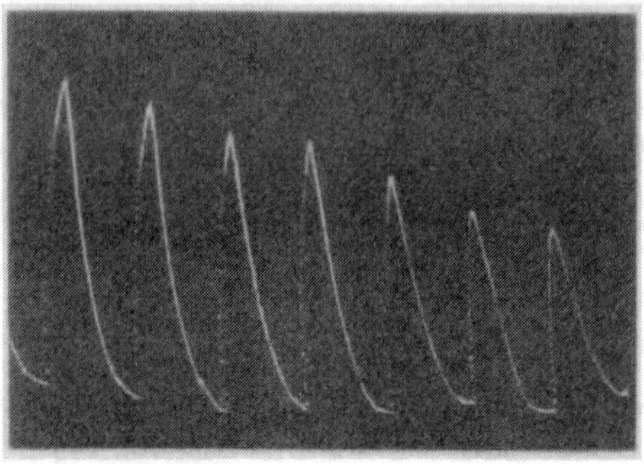

FIG. 35.—Record showing growing fatigue in *Mimosa*.

excitatory effect exhibited in the second response. As the excitation in this case was relatively slight, the recovery was very much more complete than in the first. The

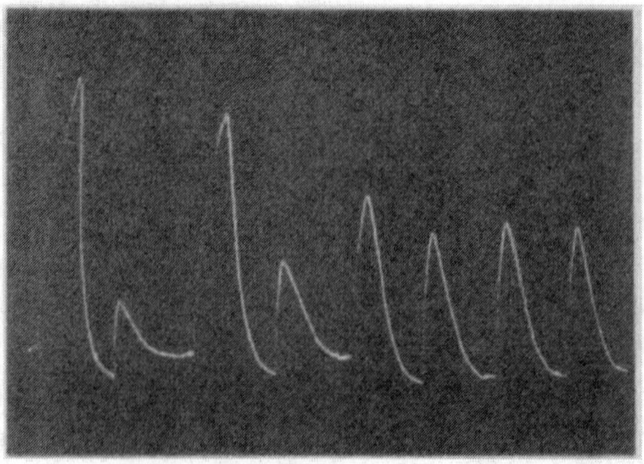

FIG. 36.—Periodic fatigue; alternate responses here tend to become uniform.

third response therefore was large, but not so large as the first, when the organ was fresh. This excitation, however, being less than the first, recovery is also somewhat more complete, and the subsequent fatigue is less than after the first response. Therefore the fourth response, though small, is not so small as the second. Thus while the first, third, and odd series of responses are progressively diminishing from a maximum, the even series—second, fourth, and so on—are increasing from a minimum. In this way the difference between the successive responses is tending to disappear, a process which is practically complete in the seventh and eighth, after which uniformity is attained. It is very interesting to note that the sum of heights of each pair of responses is approximately the same for successive pairs, and the height of a response in the uniform series is not appreciably different from the mean of the maximum and minimum of the preceding pairs, as will be seen from the following table :—

TABLE GIVING HEIGHTS OF SUCCESSIVE RESPONSES

Number.	Height of response.	Mean of successive pair.
	mm.	mm.
(1) ..	30	19
(2) ..	8	
(3) ..	26	19.2
(4) ..	12.5	
(5) ..	18.5	18
(6) ..	17.5	
(7) ..	17.5	17.2
(8) ..	17	

In this adjustment to uniformity we are able to watch a tuning of the organ, as it were, its gradual accommodation to the stimulus impinging upon it. Uniform responses may often be obtained in this way after a preliminary period of variation.

The periodic variation seen in the above cases sometimes finds still more complex expression. This is the case where waning and waxing occur in series instead of simple alternation. That is to say, response may undergo a continuous diminution in a sequence of three or more, to be followed by a corresponding sequence in which the amplitudes wax larger and larger, such serial alternations being repeated.

Staircase Response

We have seen the responses that characterise highly excitable specimens, in which there is an exhibition of growing fatigue. Taking a specimen in the contrasted condition of more or less sub-tonicity, we obtain an equally characteristic effect, which is the antithesis as it were of that which we have been considering. In this, successive responses undergo a gradual enhancement, or what is known in muscle-response—with which it is exactly parallel—as a staircase increase (figs. 37, 38). After attaining a maximum excitability, under successive stimulations, there generally ensues a fatigue-decline.

Before entering on a detailed description of this particular response it would be well to discuss certain phenomena characteristic of a relatively a-tonic condition of the tissue. In a specimen in the normal condition there is a certain amount of tonicity, accompanied by a moderate degree of contraction. When deprived of the invigorating influences of favourable external stimuli the plant becomes sub-tonic, such relative a-tonicity being characterised by relaxation or the absence of normal tonic contraction. Under the action of successive stimuli the tonic condition of the specimen will be improved. The loss of tone, with its consequent relaxation, will gradually give place to a better tone with increasing tonic contraction. Or the same improvement of tone might take the form of a gradually increasing excitability. Hence the gradual bettering the tonic condition, under successive stimulations, may often find two simultaneous expressions. In the first place the growing tone, with its increasing normal tonic contraction, will be seen in the shifting of the base-line upwards. Secondly, it will be exhibited in the growing amplitude of successive responses. These two features will

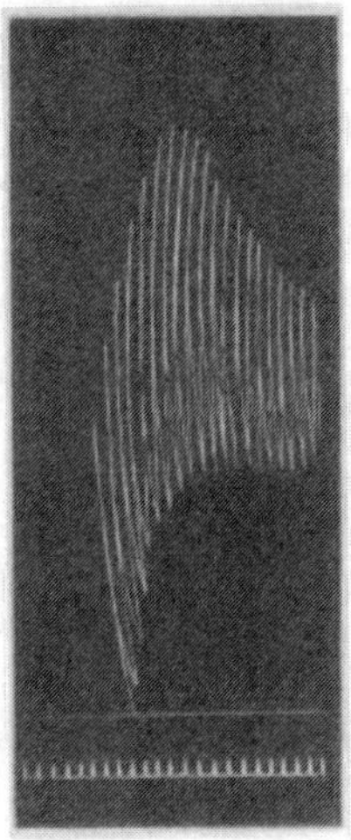

FIG. 37.—Preliminary staircase followed by fatigue in the response of frog's muscle. (Brodie.)

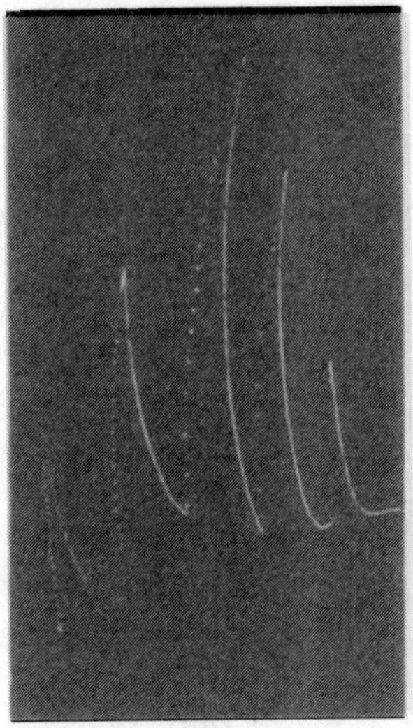

FIG. 38.—Staircase response followed by fatigue in *Mimosa*.

both be noticed in the record depicted in fig. 38. Here, as might be expected, in a specimen in sub-tonic condition we find that the first stimulus gives rise to a relatively feeble response. But in consequence of stimulation the tonic condition itself is improved, as demonstrated by the fact that the leaf remains in a slightly more contracted attitude than at the beginning. The next stimulus finds it in a better tonic condition, with accompanying higher excitability. Hence the response is larger. In this way the tonic condition reaches an optimum, with the attainment of highest degree of excitability. Here impinging stimulus has evoked the maximum response.

We see in a general way that in these responses the accession of stimulus has given rise to two kinds of effects, external and internal, whose relative values have been progressively changing. At the beginning a portion of the stimulus was utilised to improve the tonic condition, the complementary portion inducing external response. Hence at the beginning the response was small. At the end of the series, however, where the maximum tonicity has been attained, the whole blow of the stimulus is utilised in giving external response, which now therefore is maximum. After this attainment of maximum excitability the usual fatigue-decline is seen to have taken place.

We must nevertheless be on our guard against drawing too hasty a conclusion, as regards the tonic condition, from the relaxation or contraction seen in the record; we should remember that a relaxed condition is not only indicative of a-tonicity, but may also be brought about by fatigue due to over-stimulation. The changing position of the leaf, owing to daily periodicity, should also be taken into account. Bearing in mind, however, the immediately preceding history of the given plant, the experimenter will not find it difficult to guard himself against wrong inferences.

Being desirous of ascertaining how far the theoretical considerations here advanced would be borne out in extreme cases, I tested a specimen which from appearances was not at all vigorous and likely to be a-tonic. The record it gave at the beginning, of increasing relaxation, probably indicated its growing a-tonicity (fig. 39). That it was lacking in tone at once became evident from the fact that the first stimulus —applied at the point shown by the thick dot—did not evoke any response. But that this nevertheless did cause improved tonicity, is seen from the fact that the former rate of relaxation underwent a diminution, the record tending to become more horizontal. The second stimulus was then effective in evoking a feeble response. The most striking fact, however, is that on the completion of recovery the specimen actually exhibited a growing contraction as an after-effect of stimulus. Thus, while at the beginning a growing condition of a-tonicity gave rise to increasing relaxation, afterwards in consequence of stimulation this state of things became reversed, and we have a growing condition of tonic contraction appearing as the after-effect of stimulus. That the tonic condition in fact became improved is shown by the large response evoked as the immediate effect of the usual stimulation.[1]

This after-effect of a single stimulus in inducing a second contraction is significant as showing the possibility of holding incident stimulus latent for a time, to find expression later. It heralds the phenomenon of Multiple Response, which we shall consider in a subsequent chapter.

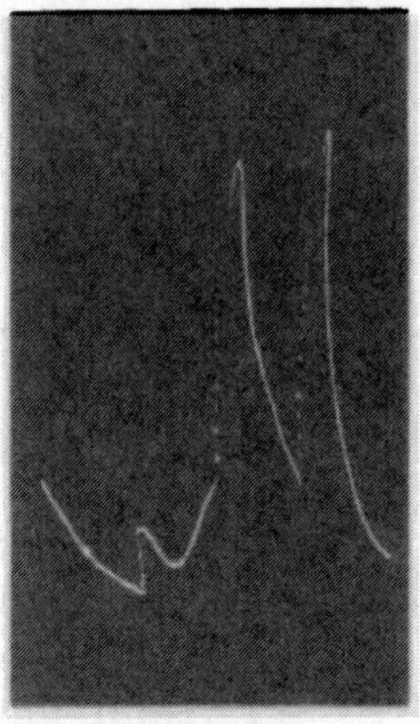

FIG. 39.—Record showing the effect of stimulus modifying tonicity, and producing staircase effect.

The curious phenomenon of alternation sometimes observed in a highly excitable specimen has already been noticed (fig. 36). The characteristic peculiarity observed there was a large response followed by a small one, such alternation continuing for a time. The difference between successive responses, however, vanished after a time. With plants in a sub-tonic condition the phenomenon of alternation is also found occasionally. The characteristics here exhibited (fig. 40) are in sharp contrast to those seen in fig. 36. Here the first response is small, and the second

[1] When the specimen is extremely sub-tonic the sign of response may even be reversed into abnormal erectile movement. After a period of stimulation, however, the response is converted into normal.

large, and the difference between the pairs in the series goes on increasing. The sum of heights of pairs of successive responses remains however approximately the same. The odd numbers in the series decline continuously (1) 18·5, (3) 16, (5) 11, (7) 6; whereas the responses in the even series grow in amplitude (2) 25·5 (4) 30, (6) 33.

TABLE GIVING HEIGHTS OF SUCCESSIVE RESPONSES

Number.	Height of response.	Sum.
	mm.	mm.
(1) ..	18.5	44
(2) ..	25.5	
(3) ..	16	46
(4) ..	30	
(5) ..	11	44
(6) ..	33	

FATIGUE-REVERSAL UNDER TETANISATION

If the *Mimosa* leaf be subjected to continuous stimulation it has been found that, after the preliminary fall, it re-erects itself in spite of the stimuli which are still acting upon it. This at first sight would appear to be very perplexing, but the apparent anomaly would however disappear when we recognise the essential unity of response in the plant and the animal. A frog's muscle, under continued tetanising electric shocks, at first exhibits the normal contraction, but afterwards relaxes, in spite of the excitation to which it is being subjected (fig. 41). The difference between the normal relaxation of recovery (expansion) and this fatigue-relaxation induced under continuous stimulation, lies in the fact that in the former case response takes place on renewed stimulation, while in the latter the tissue has become irresponsive, and only after a period of rest can it exhibit excitation. The same phenomena are observed in the case of the contractile organ of *Mimosa*. Here also, after

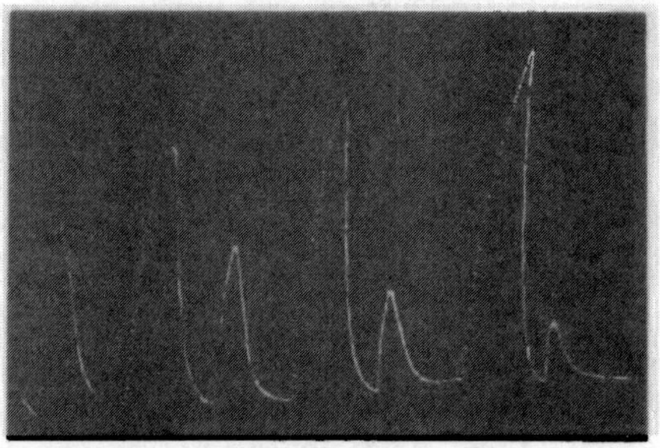

FIG. 40.—Alternating response; differences between alternate responses become accentuated.

erection (expansion) under continuous stimulation, the leaf is irresponsive and only renews its excitability after a definite period of rest.

As regards this particular reaction in *Mimosa*, we are

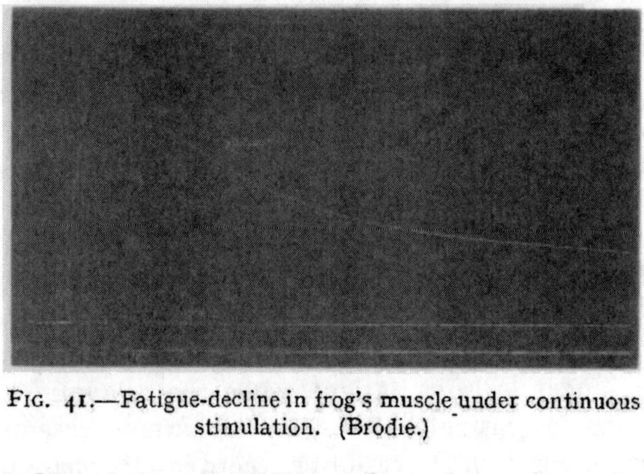

FIG. 41.—Fatigue-decline in frog's muscle under continuous stimulation. (Brodie.)

in a position to trace out the various phases through which contraction under single stimulus is reversed to expansion under fatigue induced by continuous stimulation. It has to be borne in mind that the effect of continuous stimulation is, after all, the effect of successive stimuli with the resting interval shortened. On referring back to fig. 38 we notice two phases in the response-series: in the first phase the excitability is increasing; in the second phase, it is decreasing. In the first phase again, we notice that there is a residual contraction, the recovery being incomplete. Owing to this, the base-line is gradually shifting upwards. This, coupled with the enhancing excitability and consequent staircase increase in the individual responses, brings about a maximum additive contraction, as will be understood, by joining the tops of these contractile responses. The additive effect of such contractions would be a responsive fall much greater than could have taken place under any single stimulation.

If we were now to repeat this experiment, shortening the intervals between the successive stimuli, we should obtain a somewhat similar result, with the sole difference that the successive component responses would appear nearer each other and with their recoveries still further reduced. The result of this would be slight notches in an ascending curve. Carrying this process to a limit—that is to say, when the successive stimuli follow each other quickly, as in continuous tetanisation—the notches themselves will disappear and we shall have merely an ascending curve.

Turning to the second phase in the response-series, where the excitability has reached a maximum, we find these phenomena reversed. The leaf having attained its maximum limit of fall, its capacity for further contraction is now reduced. In sharp contrast to the first phase of the series, however, successive contractions now grow smaller and smaller, under growing fatigue, while the relaxations tend to become increasingly large. In the extreme case of continuous tetanisation the resulting record in this phase would be one of relaxation, appearing as a down-curve. Thus under tetanisation we should have a response-curve, showing first the normal contraction, followed in the second place by relaxation, not at first sight very different from the response-curve due to a single stimulus. There would nevertheless be an actual difference, inasmuch as the resulting contraction under tetanisation would, on account of additive effect, be greater than that caused by a single stimulus. After the apparent recovery, due to fatigue-reversal under tetanisation, however, the excitability, as already shown, is temporarily abolished; whereas after the normal recovery from a single stimulus, excitability is fully restored.

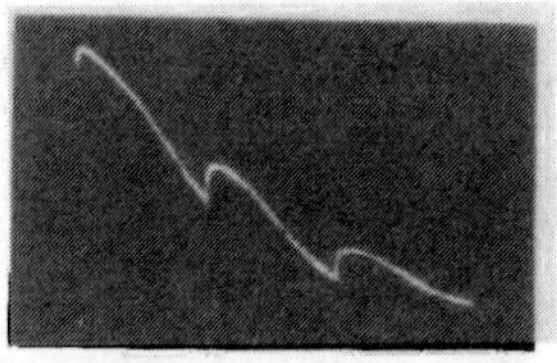

FIG. 42.—Different phases in the fatigue-reversal in plant.

The typical case, the detailed consideration of which led us to these conclusions, was that of a plant which was in a somewhat sub-tonic condition. Had the plant been in the optimum condition to start with, then following the same line of reasoning we should expect that the curve of tetanisation would be modified in a definite way. Referring back to fig. 35, which gives successive records of a highly excitable specimen, we find in this instance that the very first stimulus evoked the maximum response, and that the subsequent responses exhibited fatigue. There is not here, to begin with, any staircase effect, nor are the contractions additive, the initial response being the greatest possible. On increasing the frequency of stimulation we should, after the first maximum response, obtain the phasic variation due to fatigue. The successive contractile responses would thus appear smaller and smaller, their respective recoveries being

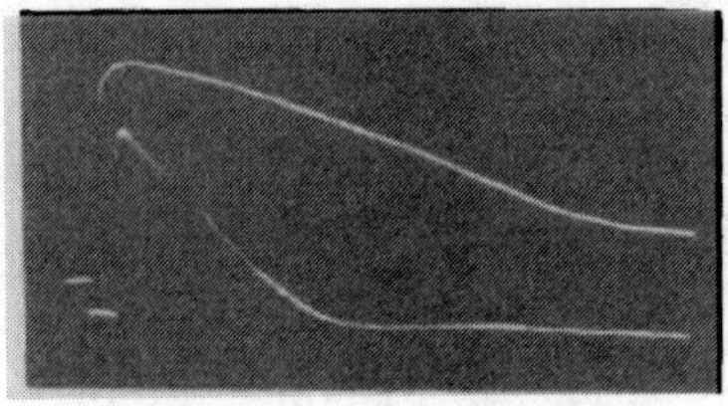

FIG. 43.—Fatigue-reversal in *Mimosa.* Lower record shows response under single stimulus; upper figure exhibits response under continuous stimulation.

correspondingly larger and larger. This is clearly seen in fig. 42, where the successive stimulations are applied at intervals of seven minutes.

Thus on subjecting a specimen in an optimum condition to continuous stimulation, we should expect to find that the extent of contraction due to tetanisation was but little different from that due to a single stimulus. This is verified by the following pair of records (fig. 43) showing the response of a plant near optimum condition, under single stimulus and under tetanisation.

Summary

The contractile response of the pulvinus of *Mimosa* exhibits characteristics similar to those of the response of muscle.

Under normal conditions of the plant, and with sufficient intervening periods of rest, the responses are found to be uniform.

The responses exhibit fatigue under conditions of incomplete recovery.

The excitability of the plant in a sub-tonic condition is enhanced by the action of the stimulus itself. Under such conditions the responses exhibit a staircase increase.

The anomalous erection, after a preliminary fall of the leaf of *Mimosa* under continuous stimulation, is explicable on the common characteristics of response in plant and animal tissues. In both, contraction is reversed to relaxation under fatigue.

CHAPTER VII

EFFECTS OF DIFFERENT GASES ON EXCITABILITY OF *MIMOSA*

Induced change of excitability under sudden variation of light—Abolition of excitability by absorption of water—Restoration of excitability by application of glycerin—Stimulating, depressing, and toxic agents—Phenomenon of accommodation—Stimulating action of ozone—Effects of: carbonic-acid gas, vapour of alcohol, ether, carbon disulphide, coal gas, chloroform, ammonia, sulphuretted hydrogen, laughing-gas, nitrogen dioxide, and sulphur dioxide.

In order to investigate the effects of various gases in modifying the excitability of *Mimosa*, a series of responses, more or less uniform, is first obtained under uniform stimuli, at intervals of 15 minutes. The given gas is now introduced into the plant-chamber, and another series of responses are once more obtained by the action of the same stimuli as before. The variation of amplitude of responses then gives an indication of the excitatory or depressing action of the agent.

In carrying out the experimental investigation in this manner, we proceed on the assumption that the stimuli applied are invariable, and that the external conditions are maintained constant, with the sole exception of the change induced by the introduction of the given gas. In order to complete a single investigation a period of nearly two hours is often necessary, which is the time required to take eight responses at intervals of 15 minutes. Of these, the first two give the normal responses, the next four the modified responses under the influence of the gas, and the last two exhibit the after-effect on the removal of the gas. It will thus be understood how important it is to maintain the external conditions constant for so long a period as two hours. The method of maintaining the testing stimulus constant has already been explained. With special care the temperature of the plant-chamber can also be kept uniform. The other factor which is liable to variation is the intensity of light. I have often noticed a fluctuation

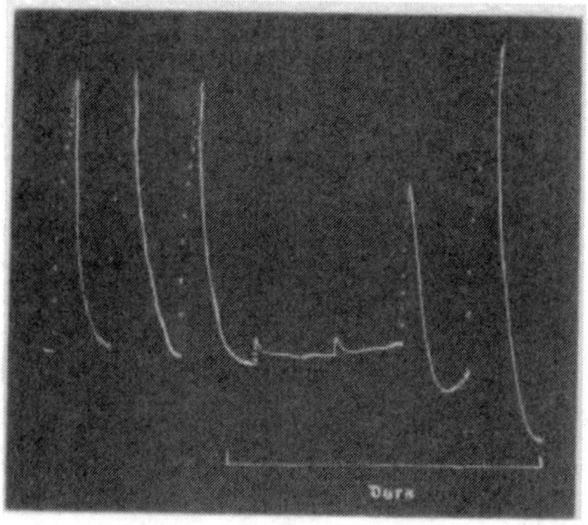

FIG. 44.—Effect of sudden darkness on excitability of *Mimosa*. First three responses, normal; four succeeding responses due to effect of darkness. Line below indicates period of darkness. Vibration frequency of writer, five times per second.

in the uniformity of responses which was traceable to a passing cloud. I soon found that a sudden change in the intensity of light induces a marked variation of motile excitability in *Mimosa*. Thus on bringing a highly sensitive plant to a dark room its excitability is found to disappear. This abolition of excitability is generally speaking temporary, since the plant often regains its sensitiveness after about an hour, though still kept in the dark. The fact that under normal conditions it is the sudden diminution of light rather than darkness that induces depression of excitability, is borne out by the fact that the plant is fully sensitive at night.

Effect of Sudden Darkness

In order to demonstrate the variation of excitability induced by sudden diminution of light, I first took a set of three normal responses in diffuse daylight. The plant-chamber was then suddenly darkened by means of an opaque screen. It will be noticed (fig. 44) that the next two responses were nearly abolished ; the excitability of the plant was however beginning to be restored after 45 minutes' exposure to darkness. After an hour in darkness the excitability was fully restored, the response here being even larger than in light.

In order to guard against the disturbing effect of variation of light it is advisable to carry out the following experiments in an open veranda, the plant being kept in a chamber with frames of ground glass. In this way the plant is maintained under diffuse light of fairly uniform intensity.

Effect of Absorption of Water

Another peculiarity I noticed in *Mimosa* was a depression of excitability on rainy days. This effect I was afterwards able to trace to the absorption of water by the pulvinus. The variation of motile excitability by absorption of water is very clearly exhibited in the accompanying record (fig. 45). A pair of normal uniform responses were first taken. A drop of water was then applied on the pulvinus, when the leaf was recovering from the second stimulus. It will be noticed that the period of recovery became very much protracted in consequence of absorption of water. The usual time for complete recovery is about 15 minutes. In the present case it was prolonged to 45 minutes. Testing stimuli were applied at the usual intervals of 15 minutes, the moments of application being represented by thick dots. It will be seen that there is an abolition of excitability, stimuli which were formerly effective becoming now quite ineffective.

I next tried to find out whether it were possible to restore the lost excitability by artificial means. Guided by the consideration that glycerin has the power of abstracting water, I applied a drop of strong glycerin to the pulvinus. It will be noted that this had the effect of quickly restoring the motile excitability of the pulvinus. The two responses

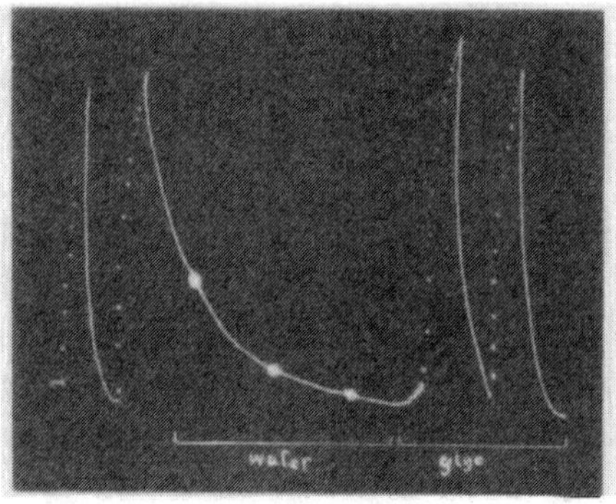

FIG. 45.—Abolition of motile excitability of pulvinus by absorption of water. Note prolongation of period of recovery and ineffectiveness of stimuli applied at moments marked with thick dots. Subsequent restoration of excitability by application of glycerin.

after the application of glycerin are practically similar to the normal responses at the beginning of the series. I am unable to say whether the restoration of excitability was here due entirely to the abstraction of water. One might think that continuous abstraction of water would induce a continuous variation of excitability—probably an enhancement reaching a maximum followed by a decline. I find, however, that the application of glycerin restores the normal excitability, and that generally speaking this remains constant even under the continued action of the reagent. This is a fortunate circumstance for those particular investigations where it is required to make an electrolytic contact with the pulvinus without inducing any change in its motile excitability.

I shall now proceed to describe the effects of various gases and vapours on the excitability of *Mimosa*. The plant is enclosed in a small glass chamber, the different gases being made to stream in and out through entrance and exit tubes. The various effects induced may be classified as (1) stimulating, (2) depressing, and (3) toxic. The exaltation of excitability induced by stimulating agents is exhibited by the enhancement of amplitude of response. The effect of depressing agents is seen in the diminution of amplitude of response; in this class may be included agents which have slight narcotic action. In all these cases the removal of the gas is attended by the restoration of normal excitability of the plant. A curious fact noticeable in this connection is the phenomenon of accommodation. Under the action of a slightly depressing agent, there is induced a diminution of excitability. But the plant may accommodate itself to the change, in consequence of which the excitability is more or less restored to the original condition. It should also be borne in mind that the character of the reaction is modified to a certain extent by the tonic condition of the plant, a plant in a vigorous condition being better able to withstand unfavourable circumstances than one in a weak condition.

Lastly, there are gaseous agents which are toxic in their action; their application is attended by rapid loss of excitability and death of the plant. I will now describe in detail the effects of various gases, beginning with those which stimulate and ending with others which cause the death of the plant.

Ozone

The stimulating effect of this gas is clearly seen in fig. 46. The particular leaf, before the application of ozone, was showing signs of fatigue, as evidenced by the gradual diminution of the heights of successive responses.

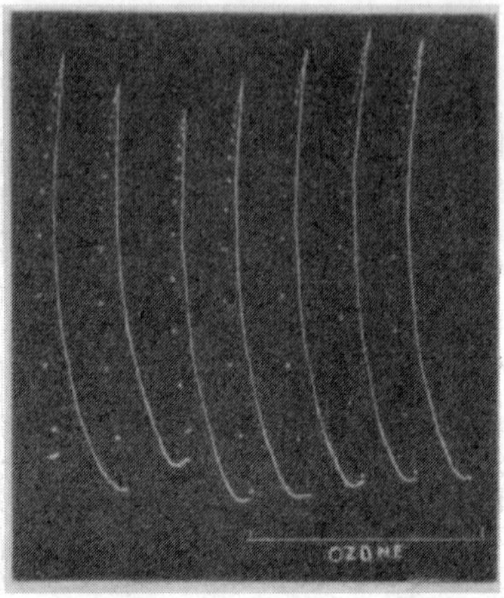

FIG. 46.—Stimulating action of ozone.

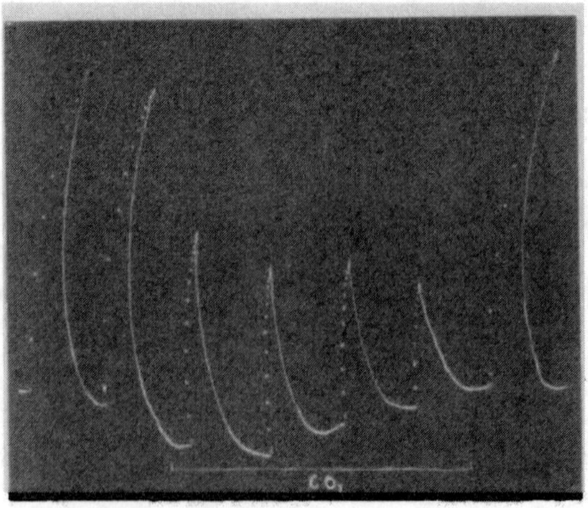

FIG. 47.—Effect of carbonic-acid gas.

The introduction of ozone brought, however, an immediate change; the induced enhancement of excitability is seen in the growing amplitude of successive responses till a limit was reached.

Carbonic-acid Gas

The effect of undiluted carbonic-acid gas is a depression of excitability. This is seen in the present record (fig. 47), where on the application of this gas the amplitudes of successive responses are seen to undergo a decline. Another noticeable fact is the incompleteness of recovery after each

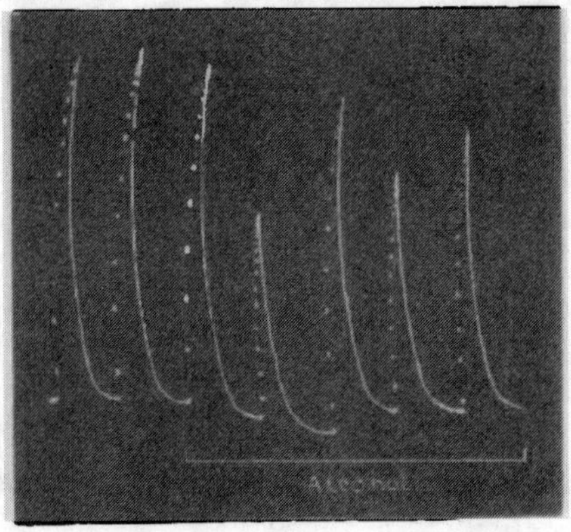

Fig. 48.—Effect of vapour of alcohol; note alternating character of response after application.

excitation. The plant-chamber was next refilled with fresh air, and we observe the restoration of normal excitability.

Vapour of Alcohol

The immediate effect of dilute vapour of alcohol is sometimes a transient enhancement of excitability. But continued action of the vapour induces a depression. In the accompanying record (fig. 48) there was little immediate effect; but after an application of 15 minutes there was induced a depression of response; another effect also noticeable is the alternating character of the response that took place after the application of alcohol.

Ether

The vapour of ether induces a depression of excitability as seen in the diminution of amplitude of response. The first effect of dilute ether-vapour is often a short-lived exaltation. The narcotic effect of this agent on *Mimosa* is feeble compared with that induced by chloroform. The

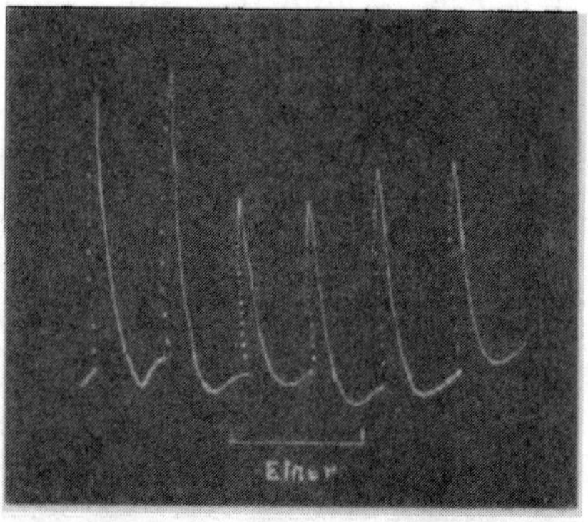

Fig. 49.—Effect of ether.

depressing effect of ether passes off on readmission of fresh air (fig. 49).

Carbon Disulphide

The effect of vapour of carbon disulphide is similar to that of ether. It induces a depression during the introduction of vapour into the plant-chamber; the induced depression, however, passes off on restoration by means of fresh air (fig. 50).

Coal Gas

Contrary to my anticipation, coal gas proved to be but moderately depressing in its action. I have kept the plant surrounded by this gas for more than two hours without the abolition of its excitability. The effect of the gas is very different when it contains impurities such as sulphuretted hydrogen. I give a record (fig. 51) which

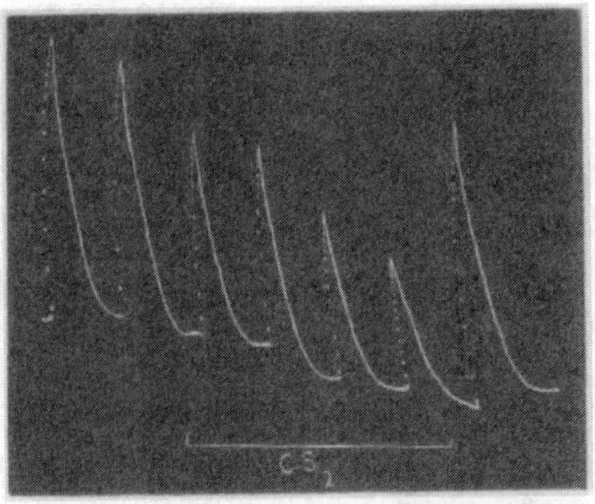

FIG. 50.—Effect of carbon disulphide.

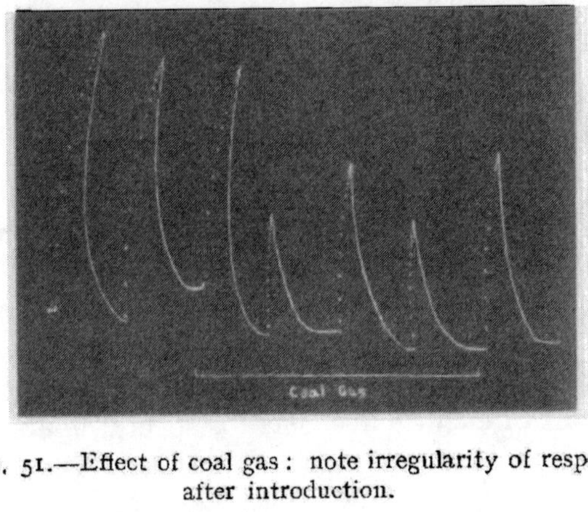

FIG. 51.—Effect of coal gas: note irregularity of response after introduction.

exhibits the depressing effect of coal gas, and the gradual restoration of normal excitability on admission of fresh air.

CHLOROFORM

The vapour of chloroform acts as a very strong narcotic. In the record here given (fig. 52) the response became very much reduced immediately after application; the power of recovery was also abolished. Subsequent application of stimulus did not result in any sign of response. Even on blowing off the vapour there was no restoration of excitability for a very considerable period. In the present case the period of total insensibility lasted for six hours, after which the excitability was slowly restored.

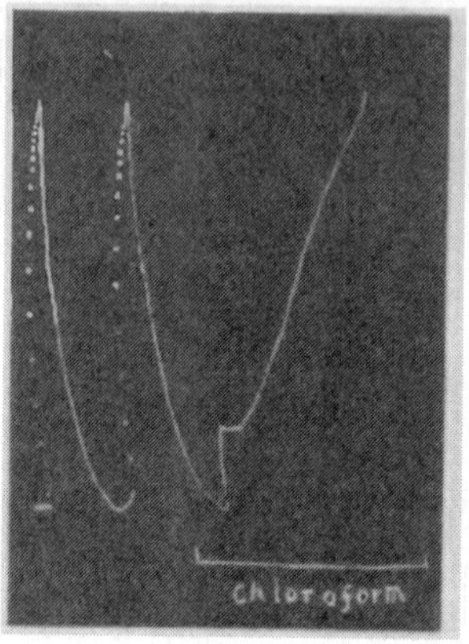

FIG. 52.—Abolition of excitability under chloroform.

AMMONIA

The vapour of ammonia is found to cause an abolition of excitability in a very short time. On the introduction of ammonia there is produced an excitatory fall. This may be avoided, however, by introducing this vapour immediately after the excitation induced by the testing stimulus. In the record here given (fig. 53), the first two are the normal responses. Introduction of ammonia is seen to induce an abolition of excitability, three successive stimulations, represented by thick dots, at the usual intervals of 15 minutes proving to be quite ineffective. On blowing off the vapour the excitability is seen to be very gradually restored. If stronger vapour of ammonia be employed, then the loss of excitability lasts for several hours.

SULPHURETTED HYDROGEN

The effect of this gas is not merely depressing but extremely toxic. It can be seen from the record that the introduction of this gas caused the period of recovery to be very protracted. The abolition of excitability is evidenced by the fact that successive stimulations at the usual interval of 15 minutes, proved to be quite ineffective (fig. 54). The action of this gas was so poisonous that restoration of fresh air did not bring about any revival. The plant was subse-

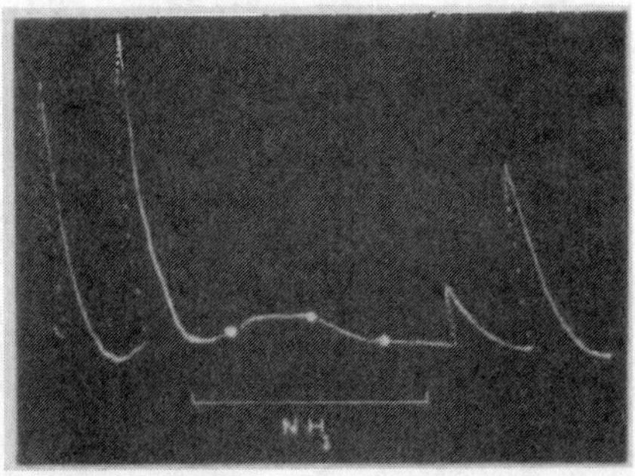

FIG. 53.—Abolition of excitability under the action of ammonia.

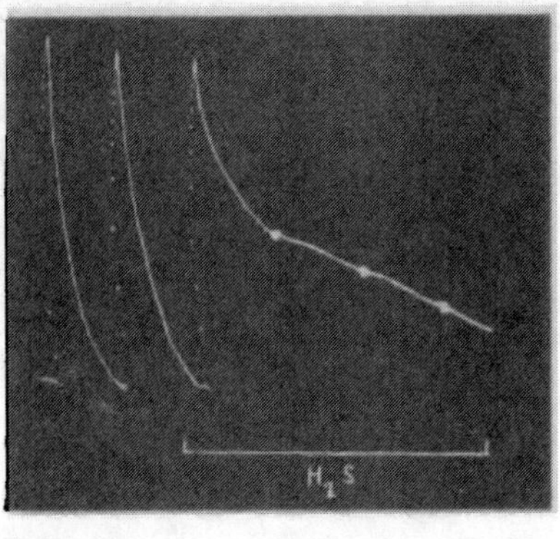

FIG. 54.—Total abolition of excitability and death of plant under the action of sulphuretted hydrogen.

quently found to have died under the poisonous effect of this gas.

NITROGEN DIOXIDE

Nitrogen monoxide or laughing-gas has but little effect: there may, however, be a slight excitatory action. Nitrogen dioxide, on the other hand, is extremely poisonous. Introduction of this gas was attended by an immediate excitatory fall, which was repeated twice. After this the plant became perfectly insensitive (fig. 55); the gas had in reality killed it.

SULPHUR DIOXIDE

Equally fatal is the effect of sulphur dioxide. Introduction of the gas was attended by an immediate excitatory

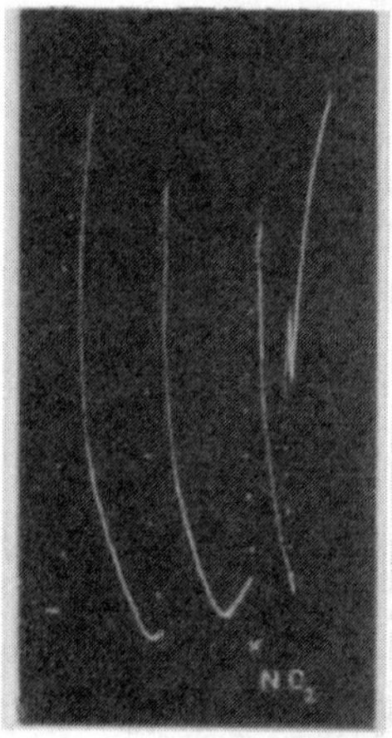

FIG. 55.—Toxic effect of nitrogen dioxide: application of gas at × induced excitation, followed by loss of excitability and death of plant.

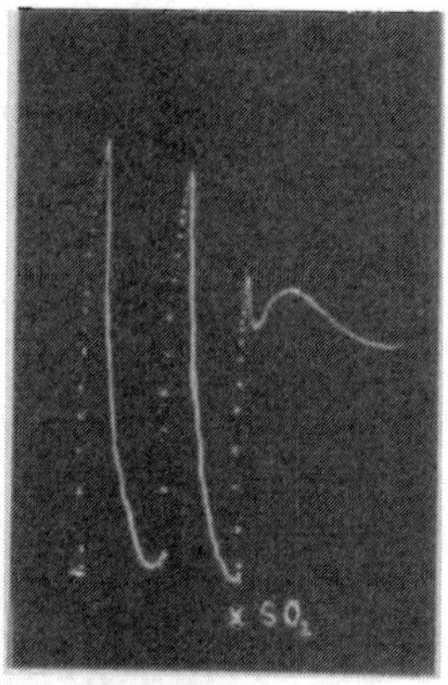

FIG. 56.—Abolition of excitability and death of plant by the action of sulphur dioxide.

fall of the leaf, after which it became quite insensitive (fig. 56). Restoration of fresh air did not revive the plant, which succumbed completely to the toxic action of the gas.

SUMMARY

There is in general a temporary depression of excitability in *Mimosa* under sudden diminution of intensity of light.

Absorption of water induces a depression or abolition

of the motile excitability of pulvinus. Excitability is restored under application of glycerin.

Ozone enhances the excitability of *Mimosa*.

Carbonic-acid gas and vapour of alcohol induce a moderate depression of excitability, which is fully restored on admission of fresh air.

Depression of excitability is also induced under the action of coal gas, and vapour of carbon disulphide.

The vapour of ether exerts a moderate narcotic action. The effect of vapour of chloroform is very pronounced, loss of excitability under its action being prolonged.

Ammonia induces a marked abolition of excitability.

Sulphuretted hydrogen, nitrogen dioxide, and sulphur dioxide abolish the excitability and bring about the death of the plant.

CHAPTER VIII

DEATH-SPASM IN PLANTS

Criterion of the death of plant—Abolition of electric response at death—Mechanical spasm of death—Water-bath for uniform rise of temperature—Excitatory effect of sudden cooling or heating—Erection of leaf with rising, and depression of leaf with falling, temperature—Thermo-mechanical inversion at the death-point—Necessity for specification of rate of rise of temperature—Death record of *Mimosa*—Abolition of response after death-spasm—Constancy of death-point exhibited by different specimens—Death records of *Desmodium gyrans* and *Vicia Fava*—Death-spasm in ordinary plants—The electric-spasm of death—Lowering of death-point by fatigue and by poisonous solution.

A PLANT may be killed by subjecting it to a certain maximum temperature. The exact moment at which death is initiated is difficult to determine, since there has been found no certain and immediate criterion of death. One method by which the occurrence of death may be determined is by the abolition of that electric response which is characteristic of the living condition. A plant as long as it is alive gives in answer to a stimulus an electric response of galvanometric negativity. On the occurrence of death this particular response disappears. I find that the electric response is abolished when the plant has been subjected for a time to a temperature of about 60° C.

If the plant is subjected to a gradual rise of temperature, there would arrive a time when the death-change will begin to occur. In the animal an early symptom of death is the setting in of *rigor mortis*. We shall find that in plants also a death-spasm, analogous to the death-throe of the animal, occurs at a critical moment.

In order to obtain an automatic record which would

indicate the beginning of death-change, I first took a specimen of *Mimosa* and subjected it to a gradual rise of temperature in a water-bath. The leaf was attached to the recording-lever in the usual manner. The recording apparatus employed was of the oscillating type, where the plate oscillates to and fro by an electro-magnetic contrivance, thus producing a series of dots in the response-curve. In the present investigation the electro-magnetic circuit is completed for a brief period, at every degree rise of temperature in the bath. Successive dots thus represent intervals of temperature of 1° C. The ordinate of the curve indicates expansive or contractile movement of the leaf : down-curve representing the expansion, and up-curve the contraction.

The temperature of the bath is continuously raised by the application of gas or spirit flame. For certain reasons, to be presently explained, it is necessary to raise the temperature *gradually* and *continuously*, without any sudden variation. There should also be no mechanical disturbance of water in the bath during heating, as that would disturb the leaf and vitiate the record. These difficulties are overcome by constructing the heating-bath of two vessels, one placed within the other. Heating the water of the outer vessel raises the temperature of the water in the inner in a very even manner, and without any mechanical disturbance.

It is necessary to subject the plant to *gradual* rise of temperature in order to protect it from excitation. Any *sudden* variation, due either to lowering or raising of temperature, causes excitatory movement of the leaf. This is seen in the following records (fig. 57), obtained with *Mimosa*. The first response is of excitation due to application of a drop of ice-cold water on the pulvinus ; the second response is due to the very opposite treatment of application of a drop of hot water. In both cases we obtain the excitatory fall of the leaf.

The effect of temperature as such is, however, very definite : *gradual rise* of temperature inducing progressive erection of the leaf; *gradual lowering* of temperature, on the other hand, inducing progressive depression of the leaf. Thus the effect of temperature, as such, is expansion with rise and contraction with fall. These opposite effects of erection and fall are progressive and slow. Excitatory reaction, on the other hand, is sudden and always attended by the contractile fall of the leaf.

The *Mimosa* used for experiment may be an entire plant; or, if more convenient, a cut branch containing a leaf may be employed. The result obtained is the same in both cases. It is found that during continuous rise of

FIG. 57.—Excitatory response of *Mimosa* induced by sudden application of either cold water (C) or hot water (H).

temperature the leaf is erected till it reaches a critical temperature at which the expansion is converted into a spasmodic excitatory contraction. The curve is thus v-shaped, the turning-point of the thermo-mechanical curve being very sharp and definite. Under constant conditions, the critical point of inversion is also very definite. The sudden inversion marks the initiation of the death-change.

Here it is necessary to bear in mind certain conditions for the securing of definite results. It is obvious that death will ensue if a plant be placed in an unfavourable environment as regards temperature for a prolonged period. But as such a temperature would only cause the death of the plant by indirect and cumulative action, it cannot be said to constitute the death-point. For precision in such a determination it is necessary to discover a temperature which is of itself efficient to initiate an abrupt death-change. On the other hand, there must be a certain latent period after the expiration of which the change would be outwardly manifested. An interval will elapse, moreover, during which the tissue is attaining the temperature of the bath. If the rate of rise of temperature be too rapid, then, owing to the lag caused by the two factors, by the time the death-spasm commences the recorded temperature may have gone beyond the actual death-point.

There are thus two points which are somewhat antagonistic. In the first place, in order to obtain the immediate point of death it is necessary that the plant should undergo an exposure which is not too prolonged. Nevertheless, to make due allowance for the latent period and for attainment of the surrounding temperature, the rate of rise of temperature must be gradual. In the case of tissues which are not too thick, the latter condition is sufficiently fulfilled by a rate of rise of 1° C. per minute. For the precise determination of the death-point the rate of rise of temperature must be specified. It must also be borne in mind that after the initiation of the death-change a certain time must elapse before the whole mass of tissue in the interior is killed. With a thick mass of tissue, owing to its inefficient thermal conductivity, the attainment of the surrounding temperature and occurrence of death throughout the tissue will be a protracted process.

The definite rate of rise of temperature may be simply secured by moving the heating flame nearer to, or further from, the bath. With thin organs, such as the pulvinus of *Mimosa*, I find that a spasmodic contraction takes place at or very near 60° C., when the rate of rise of temperature is approximately 1° C. per minute. This is seen in fig. 58; the record was commenced at 25° C., and the successive dots in the record are at intervals of 1° C. The down-curve indicates the expansive erection of leaf. As soon as the temperature had reached 60° C. there was an abrupt inversion, and the spasmodic contraction took place at a very rapid rate. The successive dots in the up-portion of the curve are at intervals of ·2 of a degree. The point of inversion, as we shall see, indicates the death-point, and the curve giving the death-record we shall call the DEATH-CURVE. It should be remembered that the particular

FIG. 58.—Death-curve of *Mimosa*. Successive dots in down or expansive part of curve represent rise of temperature of 1° C. Spasmodic contraction causing inversion of curve takes place at 60° C.

electric response characteristic of living condition of the tissue is found to disappear after the tissue had been subjected to the temperature of 60° C.

If the sudden contraction that takes place at 60° C. should prove to be the death-spasm, then this should be the last response given by the plants. If we raise the temperature of the plant short of the death-point, say to 45° C., we get a continuous responsive expansion; when cooled the leaf recovers, to a greater or less extent, its former position. If the temperature be raised again, there is once more a growing erection, and when the death-point is reached there is a sudden spasmodic contraction.

But if the specimen once passes through the temperature at which the spasm takes place, then there should be an abolition of all further response, proving the sudden contraction at 60° C. to be the spasm of death. Thus, after obtaining the sudden inversion of the curve at 60° C. in the last experiment, the plant was kept at that temperature for 15 minutes. Cold water was now substituted in the bath, and the record was taken once more of the effect of rise

FIG. 59.—Abolition of response to warming or cooling in specimen which had passed the death-point.

and fall of temperature. A record is reproduced (fig. 59) which exhibits the result. In the lower curve is shown the record of effect of rise of temperature from 45° to 65° C., and in the upper the effect of cooling from 60° to 45° C. It is seen that while in the last experiment the plant exhibited a spasmodic contraction at 60° C., there is no trace of such an effect in the present case. The very slight movement observable in the two curves is the physical effect of heating and cooling, quite negligible compared with the physiological erectile movement due to warming and the subsequent spasmodic contractile movement heralding the initiation of death-change.

In order to discover how constant is the death-point, I repeated the experiment with numerous other specimens. We have seen that the pulvinated organs present in the leaves of *Desmodium gyrans* and the bean plant (*Vicia Faba*) exhibit responsive movement under excitation. In fig. 60 is depicted the death record taken under standard conditions with the leaf of *Desmodium*. The record was commenced at 35° C.; it is seen that thermo-mechanical inversion took place at 61° C.

FIG. 60.—Record showing death-point of *Desmodium* at 61° C.

The next figure (fig. 61) shows the record with the leaf of bean plant. Here the responsive movements are very large. The inversion is seen to take place at 60° C.

The occurrence of death-spasm may also be shown by means of ordinary plants. If we take a hollow tubular organ, such as the hollow leaf-stalk of gourd or hollow flower-stalk of any other plant, cut it in the form of a spiral and subject it to the rising temperature of the bath, there is noticed at first an expansive movement of uncurling of spiral. On reaching the death-point, however, the former movement is suddenly reversed to one of curling.

Flowers like French marigold exhibit death-spasm by sudden movement of opening or closure.

By employing the electric mode of investigation I

found that there is induced an electric-spasm at the onset of death. When the temperature is rising, a given point of the plant-tissue exhibits increasing galvanometric positivity, till at the critical temperature there is a sudden electric inversion into galvanometric negativity. The

FIG. 61.—Thermo-mechanical inversion indicating death-point at 60° C. in leaf of bean plant.

electric-curve of death is of the same type as the thermo-mechanical curve. With specimens of *Musa* and *Amaranth* the death-point was found to be 59·5° C.[1]

As the death-spasm is a form of physiological response we should expect the curve of death to undergo modification under physiological variation. One such modification would lie in the translocation of the point of inversion, or the displacement of the death-point. Thus age has some influence, the death-point of very young specimens being lower than that of mature ones. I shall demonstrate

[1] Bose: *Comparative Electro-Physiology*, p. 202. Longmans, London, 1907.

the influences of other agencies, such as fatigue or poisonous drugs, in the displacement of the death-point.

I have already given a record which showed the death-point of the leaf of bean plant to be 60° C. under normal conditions. Employing a similar specimen, fatigue was induced in it by means of tetanising electric-shocks; the death record was then taken in the usual manner. It

FIG. 62.—Lowering of death-point under fatigue; death-spasm took place at 37° C.

FIG. 63.—Effect of poison in lowering the death-point.

will be seen (fig. 62) that in this particular case, on account of fatigue, the death-point was lowered from the normal 60° C. to 37° C., that is to say, by as much as 23° C. The lowering of the death-point, I find, is determined by the extent of fatigue.

In order to discover the effect of poisonous solutions on the death-point, I subjected a specimen of the bean leaf to dilute copper-sulphate solution and took its thermo-mechanical record (fig. 63). The effect of the poisonous agent is clearly demonstrated by an appropriate lowering of the death-point, in this case from the normal 60° C. to 42° C., or by 18° C.

SUMMARY

The electric response of galvanometric negativity is characteristic of the living condition of the vegetable tissue. Dead plants do not exhibit this characteristic electric-response.

When a plant is subjected for a time to a temperature of 60° C. its electric response disappears, such abolition being indicative of the death of the plant.

A leaf of *Mimosa* subjected to abrupt variation of temperature—either sudden cooling or sudden warming—exhibits excitatory reaction. But if the temperature be gradually raised, there is a progressive erectile movement of the leaf; gradual cooling induces a depression of the leaf.

When the leaf of *Mimosa* is continuously raised in temperature, then at a critical point the erectile expansive movement is suddenly converted into one of spasmodic contraction. This inversion takes place under standard conditions at or about 60° C. After this the response of the plant is permanently abolished.

Various other plants, sensitive and ordinary, exhibit this characteristic death-spasm at or about 60° C.

In taking an electric record it is found that an electric-spasm also takes place at the critical temperature, which is very near 60° C.

The death-point of the plant is lowered under physiological depression. Thus under fatigue induced by tetanising electric-shocks, the death-point was lowered from the normal 60° C. to 37° C.

Poisonous reagents also lower the death-point. In a particular case poisonous solution of copper sulphate lowered the death-point by 18° C.

CHAPTER IX

DETERMINATION OF THE LATENT PERIOD

Difficulties of accurate determination of *Latent Period*—Advantages of Resonant Recorder—Simultaneous tracings of tuning-fork exciter and Resonant Recorder — Automatic stimulation at a definite moment—Identical value of latent period in successive determinations—Accurate measurement of time-interval shorter than ·005 second—Latent period little affected by inertia of recorder—Tabular statement of value of different specimens of *Mimosa*—Effect of season on latent period.

WHEN the motile pulvinus of *Mimosa* is subjected to an exciting shock, a short time elapses between the incidence of this shock and the initiation of the responsive movement. This short interval is known as the *Latent Period.* In a responding muscle, similarly, contraction does not occur instantaneously on the application of stimulus. The latent period in this case is determined from the record of the muscle-twitch. When after the application of stimulus the muscle has not yet begun to contract, the record appears as a straight line. Then on the commencement of contraction, the recording-lever is jerked up and the curve likewise bends upwards. The length of the straight portion of the record, between a mark that represents the incidence of the shock and the flexure at the initiation of response, gives us the duration of the latent period. We have, however, to determine the time-value of this length. This is done by means of a sinuous curve drawn below the record by a tuning-fork, vibrating 100 or 200 times in a second. Experimenting in this manner, the latent period for frog's muscle has been determined at about ·01 second.

There are still several difficulties to be encountered in making this determination with any great exactness.

As the muscle-record and the time-record are separate, certain error is likely to be introduced in inferring the time-value of any point on the muscle-curve (fig. 64). This error becomes relatively serious when the total time to be measured is very small. There is, again, the difficulty of exactly determining the point of flexure which represents the beginning of mechanical response. More troublesome still is the error due to the inertia of the recording-lever. On account of this and the mechanical inertia of the responding muscle itself, the latent period thus obtained appears somewhat in excess of the true value.

FIG. 64.—Latent period of hyoglossus muscle: *aa*, moment of stimulation; *ab*, latent period. Time tracing 200 per second. (Brodie.)

In the apparatus which I employed, these difficulties have been reduced to a minimum. In the first place, the curve of response or phytogram is at the same time a chronogram. The error which might arise from an inference based on a neighbouring time-record is thus eliminated. I will later explain also the means that make it possible to determine the point of flexure, representing the beginning of the responsive movement, with relative accuracy. And lastly, the error due to the inertia of the recording part of the apparatus is reduced to a minimum by making the writing-lever excessively light. In the muscle recorders the weight of the recording-lever is about 3·5 grams. The lever which I employ weighs only ·04 gram. The recording part of my apparatus is thus nearly a hundred times lighter than that used for muscle records.

The accuracy of the time-record when made by the response recorder itself may be gauged from records giving simultaneous tracings of the exciting standard tuning-fork

(100 D.V.) and the resonant vibration in the recorder induced by it. This latter had been previously tuned to give exactly 100 double vibrations in a second. A light aluminium stylus attached to the tuning-fork traced a sinuous line on a falling plate of smoked glass. The top of the vibrating recorder was so adjusted as to make successive dots during its vibration, simultaneously with the tuning-fork tracings. It will be seen from the record (fig. 65) that, corresponding to the crest of each tuning-fork wave and slightly to its right, we have a dot. The record given represents a period of fourteen one-hundredths of a second, there being fourteen crests made by the tuning-

FIG. 65.—Simultaneous record of vibrating-recorder and 100 D.V. tuning-fork exciter.

fork time-marker, and exactly coincident with these are the fourteen dots made by the vibrating recorder. The interval between any two dots, therefore, is an accurate measurement of one-hundredth part of a second. If the plate be moving at a uniform rate, the interval between these dots will be uniform. But the accuracy of the time-measurements in the curve is independent of the rate of movement of the plate, for we calculate not by the distance but by the number of the dots. In the present figure the record was made on a plate which had been released and during its fall was acquiring increasing speed. The tuning-fork waves are thus gradually broadening out, and in exact correspondence to this the intervals between the dots are lengthening. When the phonograph motor which lets down the plate is just released, there is a short interval during which both that and the dependent plate are acquiring increasing velocity. After this the velocity becomes uniform. If this uniformity should be required throughout the record, the tracing of response may be taken during this later period only.

The mode of procedure, therefore, is first to make the recording-writer vibrate at its own definite frequency of, say, 100 times per second. The recording-plate is then released and later, when its motion has become uniform, we pass through the pulvinus an electrical stimulus of an instantaneous break-shock. There should be a mark made on the recording-plate corresponding exactly to the moment of stimulation. The horizontal record, consisting of a series of dotted points representing one-hundredth of a second, is suddenly deflected upwards on the initiation of the responsive fall of the leaf. The number of dots intervening between the mark of stimulation and this point gives us the value of the latent period for the specimen.

Stimulation cannot be effected by hand at any exact predetermined point on the record. This must be done automatically by the moving plate itself. We cannot again give an instantaneous break-shock without previously completing the primary of the Ruhmkorff coil, which causes a disturbing make-shock.

In order to avoid this the secondary electrodes, during make, should be short-circuited by means of a thick conducting-wire; the secondary shock is thus practically diverted from the plant through the path of least resistance, which is the conducting wire. All these requirements are provided for in practice by the special mechanical devices of the apparatus.

Device for Automatic Stimulation

The essential parts of the automatic arrangement by which a break-shock is given, at a predetermined point on the recording-plate, are shown in fig. 66.

The recording-plate is allowed to drop by pressing the

handle K. The winding disc is attached to the revolving axis of a phonograph motor. The disc is wound in a right-handed direction, which at the same time winds the spring

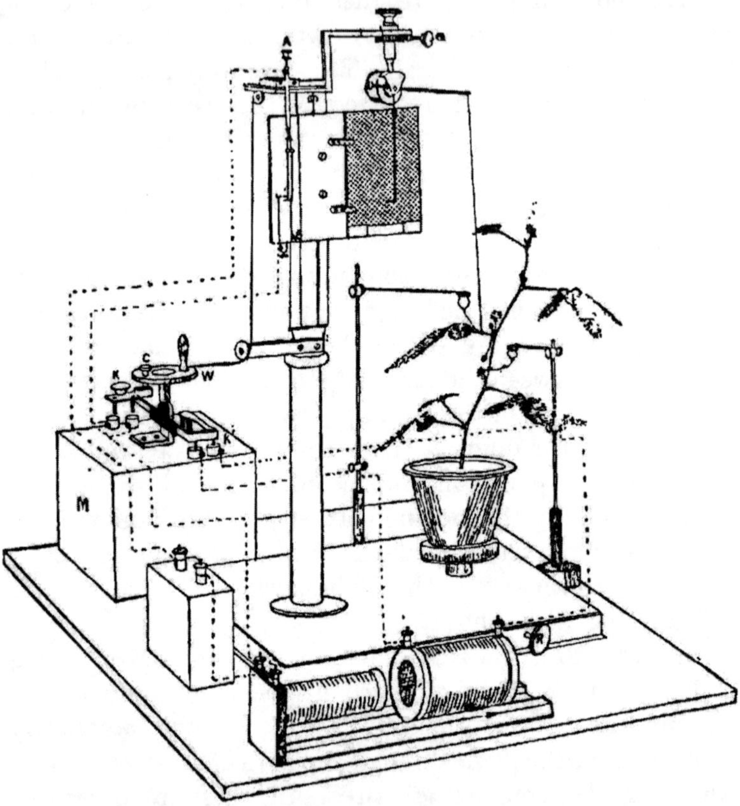

FIG. 66.—Apparatus for determination of latent period of *Mimosa*: M, spring motor; W, winding disc; C, projecting catch; K, release-handle, pressure of which also completes primary circuit of induction-coil; K', short-circuit key. The automatic break consists of contact-rod adjusted by micrometer-screw A.

of the phonograph motor. The circumference of the disc is the same as the length of the recording-plate. One complete turn pulls the recording-plate up to its highest position. A projecting catch below the disc is caught by a pin attached to the spring-handle K, when a complete turn has been made: the recording-plate is thus held arrested at its highest position. When desired, a pressure on the handle K releases the disc, the axis of the motor begins to unwind, and the plate is allowed to fall. The motor is fully wound at the beginning, and the partial unwinding during one revolution is exactly compensated before the next observation by the winding necessary to pull up the plate. Owing to the constancy of this winding, the rate of fall in successive experiments is kept the same. The pressure of the handle K, which releases the plate, also causes 'make' of the current in the primary coil. This circuit of the primary coil is completed in addition through a contact-breaking device.

This consists of a long strip of ebonite, fixed along one edge of the recording-plate carrier. On the lower end of the ebonite a conducting-strip of platinum is sunk in and provided with a binding-screw. In front of this slides a rod with contact-point tipped with platinum. This can be adjusted up or down by means of a fine micrometer-screw, A. When the recording-plate is released, carrying with it the conducting-strip, the primary circuit is broken as soon as the line of junction between platinum and ebonite is reached. This sudden interruption of the primary current gives rise in the secondary coil to an instantaneous break-shock, which passes through the plant. In order that shocks in successive experiments shall always be given at the same definite predetermined position in the fall of the plate, the following device is adopted: The recording-plate, as we have seen in a previous chapter, slides up and down a vertical support of triangular section. A movable peg fixed in the support holds it temporarily at a certain selected point chosen as that at which, during the descent of the plate, the shock is to be given automatically to the plant. For the purpose of adjustment a galvanometer is interposed in the primary circuit. So long as the point of contact-rod is in touch with the conducting-strip, so long there will be a deflection in the galvanometer. By means of its screw-adjustment, the rod is gradually raised till the line of junction between platinum and ebonite is exactly reached; the deflection in the galvanometer will now cease suddenly. In this way the point of interruption or 'break' is determined with precision. By pulling the thread in connection with one arm of the recording-lever, we then trace a slightly curved line on the smoked plate. This indicates the exact position in succeeding records of the moment of application of stimulus. This mark of stimulation is shown in the printed records as a vertical line. After making this mark on the plate, the peg is removed. It is easy to see that in successive experiments stimulation will occur at that definite moment which corresponds to this marked line of stimulation.

K′ represents a key-device by which the make-shock is prevented from exciting the plant. One end of a lever carries a bent metal-rod of U-shape, which is partly immersed in cups of mercury by means of a spring. During the depressed position of this key, the secondary coil is short-circuited. When the handle K is slightly pressed, there is a 'make' of the primary current. But the make-shock is short-circuited as K′ is still in the depressed position. Further pressure of the handle K lifts K′ up, removing the short-circuit of the secondary coil. When the break-shock is given by the contact-breaker of the falling plate, there is no short-circuit to divert the shock, which now passes through the plant and excites it.

The sequence of these events, then, is as follows:—

By turning the disc D the recording-plate is lifted and held arrested in the up-position. The pressure of the handle K releases the plate-carrier, which then begins to fall. At the same time, the primary circuit is completed and a make-shock is induced in the secondary. But this make-shock is diverted by the short-circuit key K′, which is still in the depressed position. Further pressure of the handle K removes the short-circuit by lifting the ends of K′.

All this takes place during the continuance of one pressure of the handle. During the descent of the plate, stimulation due to instantaneous break-shock takes place at the definite moment corresponding to the stimulation mark.

Electric connections are appropriately made with the plant by means of threads moistened in dilute saline solution. One electrode of the secondary coil is thus connected with the stem of the specimen: the moistened thread in connection with the other electrode is lightly wound round the pulvinus. It is sometimes preferable, for reasons previously explained, to make this contact with

FIG. 67.—Two successive records, exhibiting identity of latent period. Recorder 100 D.V. per second.

glycerin. The connections are so made that the current of the break-shock enters by the stem and leaves by the pulvinus, the latter being thus the kathode. We shall understand later the reason of this, in as much as the kathode is the point of excitation.

I now describe the record of an experiment (fig. 67) carried out in summer for the purpose of determining the latent period in a specimen of *Mimosa*. Stimulus was applied at the point marked by the vertical line, and the upper of the two records was the first taken. The vibrating-recorder employed had been tuned to exactly 100 vibrations per second; successive dots therefore represent intervals of

·01 second. It will be seen that the responsive movement begins to occur between the tenth and eleventh dots, and very near the latter. There are thus 10·9 spaces, each of the value of ·01 second, and the latent period is therefore ·109 second. In order to test to what extent successive experiments might give concordant results, I took a second record with the same specimen, which appears in fig. 67 as the lower of the two, having given the plant an interval of rest of 20 minutes after the taking of the first record. It will be seen that the second record is essentially a replica of the first, thus demonstrating that with proper precautions

FIG. 68.—Record of highly excitable specimen, taken with 100 D.V. recorder on a slowly moving plate.

successive experiments on the value of the latent period will give results which are of extraordinary constancy.

By making the travel of the recording-plate very rapid, the successive dots become more widely spaced and the minute time-intervals involved are made more conspicuous. But this has the disadvantage of rendering the flexure of the curve representing the responsive movement less abrupt, making the exact point of initiation of response somewhat more difficult to discriminate. Going, on the other hand, to the opposite extreme of making the travel of the recording-plate slow, the flexure of the curve becomes more abrupt, enabling us the better to detect the point of initiation of the responsive movement. The time-dots, however, are now closer together. This can be seen in another record (fig. 68) obtained with a vigorous specimen. Here the number of spaces before the initiation of response is eight, the latent period being therefore ·08 second. The closeness of the time-dots is not any great difficulty here, as with the help of a magnifying glass it is quite easy to make the necessary observation.

For the determination of the latent period in plants this accuracy of an order higher than hundredths of a second is more than ample. But such a limit is easily exceeded. As an example of this, I give a record (fig. 69) made with a different recorder, whose frequency was an octave higher than the last—namely, 200 double vibrations

FIG. 69.—Record of L of *Mimosa* with a 200 D.V. recorder.

each second. The successive dots are therefore in this case $\frac{1}{200}$ part of a second apart. It has been said already that by slowing the travel of the recording plate the abruptness of the flexure of the curve would be increased, the spaces between the dots being at the same time shortened. But we may obtain wider spacings without losing this sharpness of flexure, by making a magnified photographic reproduction of the curve, as shown in the next figure (fig. 70), which is a reproduction of the first part of the record in fig. 69 enlarged about three times by photographic means. In this way it is not difficult to measure, say, one-fifth of the distance between two successive dots, themselves representing an interval of $\frac{1}{200}$ part of a second. In other words, the calculation can be carried into thousandths of a second. In the present case there are

15·2 spaces between stimulus and initiation of response. The latent period of the specimen is therefore ·076 of a second.

I have been able, moreover, to construct a vibrating-recorder whose frequency is 500 times per second, a fact which enables an easy determination of time-intervals of less than a thousandth of a second to be made. These recorders, owing to their excessive lightness, possess the additional advantage of having a very small moment of inertia. It is obvious, therefore, that the employment of such recorders not only bears favourable comparison with those at present used in animal physiology, but would also

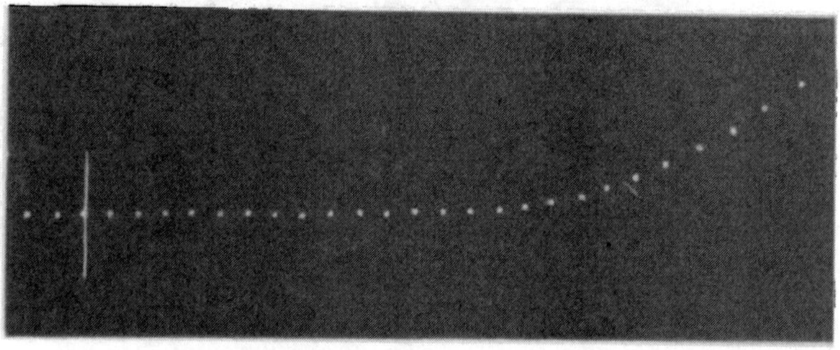

FIG. 70.—The previous record magnified.

have the advantage of reducing the error due to inertia to the lowest possible minimum, and of making the record itself its own chronogram.

It has been said that owing to the extreme lightness of the vibrating-recorder, the slight error usually due to instrumental inertia is here negligible. To what extent this is true may be judged by taking records from the same leaf with two separate recorders of different sizes and comparing the results. If the factors of inertia were prominent, then two such determinations of an identical latent period would give results varying somewhat from each other. I therefore took two different records from the same specimen, using the same stimulus but varying the mode of record—that is to say, the vibrator used in one case had been tuned to 50 vibrations per second, the length of the recorder being 12 cm. The speed of the recording-plate was in this case relatively slow. The result is shown in fig. 71. The other record in fig. 72 was taken immediately afterwards from the same specimen with a vibrator tuned to 100 double vibrations per second, the recording-plate

FIGS. 71, 72.—Two successive records taken with the same leaf; upper with 50 D.V. recorder and slow-moving plate; lower with 100 D.V. recorder and faster-moving plate. Latent period in either case is ·17 second, proving that value of L is unaffected by any peculiarity of the recorder.

moving at a faster rate. It will be seen from fig. 71 that the time-interval in the first case is represented by 8·5 spaces, each representing ·02 second, therefore proving the latent period L to be ·17 second. This, it should be mentioned, was an autumn specimen, in which the latent period is somewhat longer than in summer. In the second record, fig. 72, under its different speed and with the vibrator giving 100 vibrations per second—we find the intervening spaces to be 17. This gives the latent period as again

·17 second. The identity of these values shows that the inertia of the recorder has but little effect on the results obtained.

The latent period in any given specimen of the pulvinus of *Mimosa* is, as we have seen, under uniform conditions extremely constant. It differs, however, in different specimens and from season to season. A thin specimen has in general a shorter latent period than one which is stouter. Perhaps this fact is illustrated, with a certain exaggeration, in the case of *Neptunia*, the leaf and pulvinus of which are comparatively thick. In any case, we have already seen

FIG. 73.—Record of latent period of *Neptunia* with 10 D.V. recorder.

that in its responsive movements, relatively to *Mimosa pudica*, it is very sluggish. In order to determine the latent period I employed a slow vibrator, that is to say, one which vibrates with a frequency of 10 per second. It will be seen by reference to fig. 73 that the responsive movement began after the sixth dot, the latent period being thus ·6 second, or six times the value of the average latent period in *Mimosa*.

With *Mimosa pudica* I have carried out more than a hundred different determinations, and give below a tabular statement of seventy of those values which occurred most frequently amongst these. Specimens giving a latent period shorter than ·08 second or longer than ·12 second in summer may be regarded as rather exceptional.

TABULAR STATEMENT OF VALUES OF LATENT PERIOD L IN NUMBERS OF DIFFERENT SPECIMENS OF *Mimosa*.

Number of specimens			Value of L
10	..	..	·08″
9	..	..	·09″
26	..	..	·10″
10	..	..	·11″
15	..	..	·12″

The shortest latent period that I have come across is ·06 second obtained in summer and when the temperature was specially high. The longest in summer was ·14 second. Most specimens have a latent period not appreciably differing from ·1 second. This may be regarded as approximately the average value for summer. In winter and with sluggish specimens the latent period may be prolonged to a value of something like twice as much, that is to say, ·2 second, more or less.

SUMMARY

Latent period of *Mimosa* may be determined with great accuracy by means of Resonant Recorder. This enables the measurement of time-interval shorter than ·005 second.

Error due to inertia is reduced to a minimum on account of extreme lightness of plant-recorder, which is nearly a hundred times lighter than muscle-recorder.

Successive values of latent period with the same specimen are found to be constant. The results are not modified by employment of different recorders.

The shortest value of latent period given by a vigorous *Mimosa* leaf in summer is ·06 second, the average value being ·1 second.

The latent period of leaf of *Neptunia oleracea* is ·6 second.

CHAPTER X

INFLUENCE OF INTENSITY OF STIMULUS, FATIGUE, AND TEMPERATURE ON THE LATENT PERIOD

Diffuse stimulation under alternating shock—Effect of intensity of stimulus on *Latent Period*—Influence of optimum condition—Effect of fatigue—Effect of temperature.

I WILL now describe different experiments carried out for the purpose of observing the effect of varying external conditions—such as the intensity of stimulus, fatigue, and temperature—on the latent period. The mode of procedure adopted was first to take a record giving the latent period under standard conditions, and then to make further records under conditions similar in all respects to the first, except in regard to the one special factor whose influence was to be determined.

As some of the experiments in question necessitated a long period of observation, lasting sometimes over an hour, it became necessary to eliminate one source of possible uncertainty—namely, the effect of electrolytic contact on the pulvinus. It has been shown that there is no variation of excitability induced in the pulvinus where the contact is made with glycerin. It would, however, be preferable to effect direct stimulation without placing either of the electrodes on the pulvinus. In connection with this it was found that if one of the two electrodes were placed on the petiole slightly to the right of the pulvinus, and the other on the stem immediately below it, and a few rapidly alternating shocks passed through, excitation was simultaneous throughout the interposed tract.

This may be demonstrated in a striking manner by

experimenting with sub-petioles of *Mimosa* bearing numerous pairs of sensitive leaflets. The two electrical connections are made on the middle points of two neighbouring sub-petioles. If a single induction-shock be passed, then it will be found that the point by which the current leaves the petiole—the kathode—becomes the seat of excitation, which is transmitted serially to the neighbouring leaflets. The characteristics of this phenomenon will be dealt with in detail in a subsequent chapter. If instead of a single shock a few alternating-shocks of moderately strong intensity be next passed in rapid succession through the sub-petiole, it will be found that the excitation has become diffuse, the leaflets in the intrapolar tract exhibiting excitation simultaneosly.

This fact of simultaneous excitation in an interposed tract may be demonstrated by the following experiment giving quantitative results: Two successive records are taken of the response of the pulvinus of *Mimosa*, the exciting electrode being first placed with the interposed pulvinus 10 mm. apart and again 80 mm. apart. The distance of the pulvinus, in the first case, would be 5 mm. from either electrode, and in the second case, 40 mm. The average velocity of transmission of excitation, as will be seen in the next chapter, may be taken as approximately 16 mm. per second in summer. If the excitation in the interposed tract is not simultaneous, but locally initiated at the points of application of the electrodes, we may expect to find that the periods intervening between the beginning of stimulation and the initiation of response will differ greatly from each other in the two cases. In the first case, where either electrode is distant from the pulvinus by 5 mm., the delay in the response may be expected to be of the order of ·3 second. In the second case, where either electrode is distant from the pulvinus by 40 mm., we may expect the delay caused by transmission to be about 2·5 seconds.

If the excitation were to prove simultaneous, however,

we should find no difference of time as between the two cases; and this stimulation would be equivalent to direct stimulation and be expected to give us a latent period of the order of ·1 second.

I give below (fig. 74) the record of this experiment, which shows that under alternating shocks the excitation is simultaneous, and that the value of the latent period is then independent of the points of application of the electrodes, provided the pulvinus be included in the tract. The upper of the two records was taken when the electrodes were 10 mm. and the lower when they were 80 mm. apart. It

FIG. 74.—Simultaneous excitation in interposed tract under alternating shock. In upper record two electrodes were 1 cm. apart, in lower record, 8 cm. apart.

is seen that the latent periods obtained from the two experiments are the same—namely, ·08 second—which is of the order of other determinations already obtained. Had the stimulation not been simultaneous this value would have been increased to at least ·3 second or to 2·5 seconds in the respective instances. The identity of results in the two cases shows, moreover, that we are measuring the effect of a constant factor, which is the latent period.

For the purpose of applying alternating shocks I employed as usual a Ruhmkorff coil. In this the spring vibrator in the primary was so adjusted as to cause 100 interruptions per second. This would give in the secondary circuit 200 alternating shocks per second, of which 100 would be due to make and the alternating 100 to break.

In order to subject the plant to brief alternating shocks applied at definite moments, and having definite duration, the sliding interrupter and its connections already described were appropriately modified. The vertical sliding contact-plate now consists of a conducting platinum sheet, except for a narrow non-conducting interruption made of a piece of inlaid ebonite. The sliding-plate and the contact rod now short-circuit the secondary, except during the brief interval when the narrow strip of ebonite removes the short-circuit. It is during this definite interval that the plant is subjected to the alternating shock. The breadth of the strip is so chosen that the duration of the shock is about ·05 second. During this time the plant will receive 10 alternating shocks, 5 of make and 5 of break. If desired, a simple device may be introduced by which the duration of the shock can be modified. This modification consists in inlaying a right-angled triangle instead of a linear piece of ebonite on the metallic plate. The base of the right-angled triangle is kept horizontal. It follows that, by adjusting the rod from right to left, the interval of the removal of short-circuit can undergo continuous reduction of duration.

The mark which indicates the *beginning* of stimulation is made in the usual manner, and in successive experiments the stimulation is initiated at this precise moment. The duration of stimulus is also constant. In the following experiments, it is to be remembered, we are to determine the effect of changing one factor whilst maintaining others constant. Thus we have in each case to take one record under standard and one or more under modified conditions.

Effect of Intensity of Stimulus

The series of experiments to be described below have in each case been repeated at least twelve times, with results that were invariably concordant. I content myself, however, with giving two records in each case, obtained from different specimens. In order to test still further the reliability of these results I was careful, with each pair of figures given for comparison, to employ two different recorders, the vibration-frequency of the first being 100, and of the second or companion-set 50 per second.

In fig. 75, with vibrating recorder of 100, are given two records testing the effect of intensity of stimulus on the latent period. The lower of the two was obtained with the minimum stimulus of 1; and the latent period is seen to be ·155 second. In the upper of the two records we

FIGS. 75, 76.—Effect of intensity of stimulus on latent period: the upper record in each is due to stronger stimulus.

have the result of the maximal stimulus of 5. The latent period is now found to be reduced to ·1 second.

In fig. 76, with vibrating recorder of 50 and taking a different specimen, we find a precisely similar result. The lower of the two records, with the minimal stimulus of 1, shows a latent period of ·14 second. The upper, with stimulus 2, which in this individual case was maximal, shows a latent period of ·09 second.

It is interesting to note alike in figs. 75 and 76 the great vigour of the responsive movement under higher intensity of stimulus, as seen in the abruptness of the rise of the curve and the wider spacing of the successive dots.

The following table shows the effect of intensity of stimulus on the latent period :—

TABLE I.—EFFECT OF INTENSITY OF STIMULUS ON LATENT PERIOD.

Number.	Intensity of Stimulus.	Latent period.
1 ..	1 5	·155 ·10
2 ..	1 2	·14 ·09
3 ..	1 4	·15 ·11

FIG. 77.—Constancy of latent period when stimulus is above maximal. Lower record under stimulus 2; upper record under stimulus 5. Recorder 100 D.V.

It is thus seen that with the increase of the intensity of stimulus there is a corresponding reduction of the latent period. But it would appear from further experiments that a limit is soon reached, when the stimulus begins to be maximal. A further increase of the intensity of stimulation above this point will have little or no effect in reducing the latent period. This is shown in fig. 77, which gives a pair of records taken with a vibrator having a frequency of 100 double vibrations per second. The lower of these two records represents the effect of a stimulus of 2, which was here maximal. The upper was taken with the increased intensity of stimulus of 5. In the two cases the latent period was practically the same—namely, ·12 second.

It should be mentioned here that in a plant in optimum condition the latent period differs very little under strong or feeble stimulus. We have also seen, it will be remembered, that in an optimum condition of the specimen there is very little difference in the amplitude of response under strong and feeble stimulus respectively.

Effect of Fatigue

It has been shown that the successive values of the latent period become constant provided a resting-interval be

Figs. 78, 79.—Effect of fatigue.

allowed for complete protoplasmic recovery. The period required for full recovery I find to be about 20 to 25 minutes in summer, more or less. If this resting-interval be shortened, the effect of fatigue is seen in the prolongation of the latent period; if this shortening be carried too far, then the motile excitability is temporarily abolished. I give below a pair of records which exhibit the prolongation of latent period on account of fatigue.

The mode of procedure is first to obtain the normal record with a fresh specimen under a maximal stimulus of intensity 3. This is the intensity which is always used unless the contrary be stated. In order to exhibit the effect of fatigue, the second record is taken after a period of rest of only 15 minutes. The upper record (vibration-frequency 100) gives the normal value of L to be ·11 second; the lower record shows that the latent period has been prolonged to ·16 second on account of fatigue (fig. 78).

The experiment was repeated with a different specimen and with a vibrating recorder giving 50 vibrations per second. The record (fig. 79) shows again that the latent period is prolonged under fatigue, from the normal ·1 second to ·14 second. The effect of fatigue is independently seen in the record, in the relative sluggishness of the responsive movement. The slope in the response of the fresh specimen is almost vertical, with successive dots very widely spaced. In the response of the fatigued specimen a great contrast is observed in both these respects.

I give below a tabular statement showing results of different experiments on the effect of fatigue:—

TABLE II.—EFFECT OF FATIGUE ON THE LATENT PERIOD.

Number.	L in fresh specimen.	L′ when fatigued.
1	·11″	·16″
2	·10″	·14″
3	·10″	·22″
4	·11″	·17″
5	·8″	·13″
6	·11″	·15″

Effect of Temperature

The effect of temperature on the latent period is shown in the next two sets of records (figs. 80, 81). In fig. 80 we have three sets of records, taken with a 100 D.V. recorder at the three different temperatures of 23° C., 28° C., and 33° C. respectively, under a uniform stimulus-intensity of 2. These temperatures were maintained by means of the thermal chamber, heated electrically. From the lowest record at a temperature of 23° C. the latent period is seen to be ·165 second. At 28° C. in the middle record, it is found

Figs. 80, 81.—Effect of temperature.

to be reduced to ·125 second. And at 33° C. it becomes still further reduced to ·065 second.

In fig. 81 these results are corroborated by records taken with a different specimen, under stimulus-intensity of 2, the vibration-frequency of recorder being 50 D.V. The three records are for temperatures of 24° C., 29° C., and 33° C. respectively. The shortening of the latent period with rising temperature is also shown here in a very striking manner. The lowest of the records, taken at 24° C., gives us a latent period of ·14 second. The next, at 29° C., shows a reduction to ·102 second. And the last and highest, at 33° C., gives us a latent period of only ·07 second.

The increase of vigour in the responsive movement under rising temperature is also very clearly apparent in the record. It will thus be seen that the latent period decreases with rising temperature.

The following table gives the results of several experiments on the effect of temperature on the latent period :—

TABLE III.—EFFECT OF TEMPERATURE ON THE LATENT PERIOD.

Number.	Temperature.	Latent period.
1 ..	23°	·165″
	28°	·125″
	33°	·065″
2 ..	24°	·140″
	29°	·102″
	33°	·070″
3 ..	20°	·190″
	25°	·10″
	31°	·08″

SUMMARY

Alternating electric shocks of moderate intensity induce simultaneous excitation throughout the interposed tract.

Latent period is in general shorter under stronger intensity of stimulus. There is no further variation above a maximal.

In the optimum condition of the plant the latent period is the same for feeble and strong stimulus.

Fatigue prolongs the latent period.

The latent period is shortened by a rise of temperature.

CHAPTER XI

VELOCITY OF TRANSMITTED IMPULSE IN PLANTS

Detection of transmitted excitation by means of electromotive variation—Specific tissue for conduction of excitation—Hydro-mechanical theory of transmission of stimulus—Propagation of excitatory protoplasmic change—Physiological test—Automatic record of transmission-period—Conditions for obtaining constant velocity—Determination of velocity of transmission in *Mimosa*—Differential method of determining velocity—Constancy of results—Tabular statement of different determinations of velocity—Effect of intensity of stimulus on velocity of transmission—Effects on sub-tonic tissue and on tissue in optimum condition—After-effect of stimulus in enhancing conductivity—Effect of optimum condition—Disturbing action of leakage of exciting current—Effect of fatigue—Effect of temperature—Velocity of transmission in *Biophytum* and *Averrhoa*—Direction of preferential conduction.

WE have hitherto dealt with the reaction of tissues which exhibit the excitatory condition by motile response, as in pulvinus in the case of the plant and muscle in the case of animal. In the animal, again, we meet with certain conducting-tissues in which the excitatory protoplasmic change is transmitted to a distance, and, should one of these nerves happen to lead to a contractile muscle, the transmission of the excitatory change is conspicuously exhibited by the contraction of the terminal organ.

We now come to the question whether there is a transmission of a true excitatory change in the plant, and if so whether there is in it any specific conducting-tissue, corresponding to the nerve of the animal, for the conveyance of excitation? Since the transmission of excitation depends on the propagation of a protoplasmic change, it follows that a conducting-tissue must be characterised by a more or less protoplasmic continuity. Should the plant possess any tissue analogous to the nerve, then it is in the fibro-vascular bundle that we must look for it.

In a nerve-and-muscle preparation the transmitted excitation is detected by the contraction of the terminal muscle. Even in the absence of any terminal contractile organ, we can detect the passage of excitation by an electrical method. It is known that the excitation of a living tissue is attended by a concomitant electrical change of galvanometric negativity. If we make suitable galvanometric connections with two points on a nerve, and we stimulate the nerve at a distant point, we shall find that the arrival of excitation from the distant stimulated point is at a proper moment signalised in the galvanometer by a deflection of a definite sign.

Similarly, I have found that the excitatory change of galvanometric negativity is transmitted to a distance through certain plant-organs. Tissues containing fibro-vascular elements, such as stems and petioles, are found to be good conductors of excitation. Indifferent tissues in leaves and tubers possess little power of conduction ; in such cases excitation remains more or less localised. The parenchyma in the leaf is thus an indifferent conductor, whereas the midrib and veins are good conductors of excitation. In stems also great difference is found, as regards power of conduction, between the fibro-vascular strands and the ground tissue. The results of electrical investigation thus give strong support to the conclusion that plants possess conducting-tissues by means of which the excitatory state may be transmitted to a distance.

The prevailing opinion, however, up to the present has been that in plants like *Mimosa* there is merely a transmission of hydro-mechanical disturbance and no transmission of true excitation comparable with that of animal nerve. That this conclusion is erroneous will be shown from the results of varous inquiries fully described in the next chapter. In all these investigations it is necessary to determine the velocity of transmission with the highest accuracy ; and in order to eliminate the errors that might be inherent in personal observation, it is desirable that all the data for this determination should be furnished automatically in records made by the plant itself. Successive records, therefore, should enable us to determine with equal accuracy not only the normal velocity but also its variation under given changed conditions.

And here the preliminary questions arise: With what degree of accuracy can we determine the normal velocity of transmission? And how far may we depend on the constancy of this velocity, in successive experiments, under normal conditions? As regards these points, some misgivings might naturally arise. For the factors calculated to interfere with this constancy will in all probability prove to be numerous. First, we may have the variation of excitability at the point of application induced by the stimulus itself. We have, therefore, to find out what is the maximum intensity of stimulus that may be employed without causing fatigue or other deleterious changes in the tissue. Another point to be remembered is the question already discussed in previous chapters of our ability to apply stimuli, in successive experiments, of identical intensity and duration. Unless this can be secured we cannot look for consistent results, inasmuch as the velocity of transmission may to some extent be dependent on the intensity of stimulus. Likewise, if the transmission of excitation should prove to be due to the transmission of a protoplasmic change, it is easy to see that we must allow the tissue a definite time for protoplasmic recovery after each application of stimulus, without which interval consistency of results could hardly be expected.

It was only after a long course of investigations—some of which will be described in the course of the present chapter—that I was able to analyse and provide against these several sources of variation. But even after this, I was by no means prepared for the very great consistency of the results which it has been my good fortune to obtain.

For successive determinations, with the same specimen, of the periods required for the transmission of excitation through a given length of conducting-tissue, did not differ from each other by so much as one-twentieth of a second and were often actually identical.

Determination of Velocity of Transmission

For the purpose of these experiments I used by preference the petiole of *Mimosa*, for the reason that in this the conducting-strands situated in the fibro-vascular bundle would be more continuous and evenly distributed than in a branching specimen. In order to determine the velocity of transmission, the stimulus of induction-shock is applied to the petiole at a distance d from the responding pulvinus. Let us suppose t to be the true time taken by the excitation to reach the pulvinus; the initiation of the responsive movement will however be further delayed by the latent period of the pulvinus L. The total time-interval T observed to elapse between the application of stimulus and the initiation of response will therefore be the true time t plus the latent period L. To obtain the true time we have to subtract the latent period L from the observed interval T, thus $t = T - L$. The velocity of transmission is then found by dividing the distance by the true time. The necessary data are therefore the distance between the stimulated point and the pulvinus, the time-interval between the application of stimulus and the initiation of response, and the latent period of the individual pulvinus.

In making these determinations the apparatus employed is the same as that for the determination of the latent period. As in these experiments we have to measure time which may be several seconds in duration, the recording-plate is made to travel at the relatively slow rate of 2 cm. each second or thereabouts. The vibrating recorder must be selected according to the degree of accuracy that is required. For our present purpose a time-measurement accurate to one-tenth or one-twentieth of a second is ample.

We first obtain a series of records of indirect stimulation. The two electrodes, E and E′, in connection with the exciting secondary coil, are applied on the petiole about 10 mm. apart, the proximal electrode E being at a distance *d* from the pulvinus. The recording-plate during the course of its descent completes the primary circuit of the induction coil for a definite length of time, which is about one-twentieth of a second. This gives rise to a definite number of alternating shocks to the plant. The stimulus is always applied at a definite instant in the descent of the plate; hence successive records on the same plate always commence on the same level, the vertical line in the record indicating the moment of the application of stimulus. After taking one or more records of the effect of indirect stimulation, an additional record is taken of the effect of direct stimulation. This gives the latent period L of the particular specimen.

Before proceeding further I must point out the necessity of special precautions for the perfect insulation of the electrodes in connection with the secondary coil. If one of these should happen to touch the table, then, even with connections made for indirect stimulation, a portion of the current would pass through the flower-pot holding the plant and the pulvinus would be directly stimulated by this escaping current or current of leakage. In my own case, it was some time before I discovered that certain anomalous results were to be traced to this particular source of disturbance, at first little suspected. To overcome this difficulty the flower-pot should be placed on a block of insulating ebonite, the electrodes also being carefully insulated on ebonite rods.

I will now proceed to give the actual records obtained with the arrangements detailed above. In the experiment I am about to describe the specimen of *Mimosa* was very vigorous. The distance at which the stimulus was applied was 30 mm. from the responding pulvinus, and the intensity of stimulus was 3 units. The frequency of the vibrating-recorder was 10 per second ; hence the distance between any two successive dots in the record represents a time-interval of one-tenth of a second, and from the record itself it will not be found difficult to estimate intervals of even one-fifth that amount. The lowest of the three records in fig. 82 represents the results of the first experiment. It will be seen that the interval between the stimulus and the

FIG. 82.—Determination of velocity of transmission of excitation in *Mimosa*. Two lower records are in response to stimulus applied at a distance of 30 mm. ; upper record in response to direct stimulation giving the latent period. Recorder 10 D.V.

beginning of response is 16·2 spaces, each of the value of ·1 second. Therefore the total time T is 1·62 second. After a suitable interval necessary for complete recovery of conductivity, a second record was taken, under the same conditions, on the same plate. The minimum interval necessary varies from 15 to 20 minutes, depending on the condition of the specimen and the season. It will be seen that the time-interval in this case is the same as before—namely, 1·62 second. The third record was taken with direct stimulation, and from this it will be noted that the latent period is ·12 second. Thus the velocity of transmission as given by both these experiments is identical—namely,

$$t = 1{\cdot}62'' - {\cdot}12'' = 1{\cdot}5''$$

$$V = \frac{d}{t} = \frac{30}{1{\cdot}5} = 20 \text{ mm. per second.}$$

The Differential Method

In order to put the constancy of these results to a still more rigorous test, I next modified the experiment in the following way, employing the *Differential Method.* The stimulus was first applied at a distance d from the responding pulvinus, and the total time T was found from this record. In the next experiment the distance of the point of stimulation was reduced to d_1, and its corresponding total time, T_1, found in the usual manner. And lastly, a record was taken under direct stimulation. This furnished the value of the latent period L.

It will be seen that we have here three different sets of data for the determination of the absolute value of the velocity of transmission. In two of these we derive the velocity in the usual manner from the distance, the total time, and the latent period. In the third, knowledge of L is not required, as it will be seen that the difference in the times $T—T_1$ of transmission observed in the first two cases represents the time taken by the excitation to travel the difference between the two distances $d—d_1$. Hence three separate determinations, V_1, V_2, and V_3, obtained with the same specimen, are

$$V_1 = \frac{d}{T-L}; \quad V_2 = \frac{d_1}{T_1-L}; \quad V_3 = \frac{d-d_1}{T-T_1}$$

The rigour of this test of constancy will be gauged by the extent to which the different determinations of V_1, V_2, and V_3 are consistent with one another.

In an experiment carried out in this way the intensity of stimulus applied was 3 units, and the vibrating recorder had a vibration-frequency of 10 per second. In the first experiment the point of application of stimulus was at a distance of 30 mm.; the total time was found to be 1·9 second. In the next experiment the distance was reduced to half, that is to say, 15 mm., and the total time was found to be 1 second. And lastly, the latent period was determined, under direct stimulation, at ·08 second (fig. 83). Thus

$$V_1 = \frac{30}{1{\cdot}9 - {\cdot}08} = 16{\cdot}4 \text{ mm. per second}$$

$$V_2 = \frac{15}{1 - {\cdot}08} = 16{\cdot}3 \text{ mm. per second}$$

$$V_3 = \frac{15}{1{\cdot}9 - 1} = 16{\cdot}6 \text{ mm. per second}$$

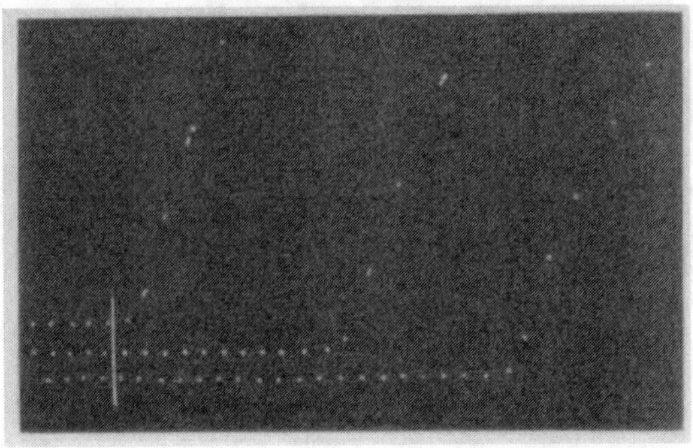

FIG. 83.—Determination of velocity by differential method. The records from below to above are in response to stimuli applied at distances of 30 mm., 15 mm., and directly. Recorder 10 D.V.

The three results thus obtained from independent data are here seen to be extremely consistent. They bear very emphatic testimony not only to the accuracy of the method but also to the constancy of the velocity in a given specimen under unvarying external conditions.

It has been stated that the accuracy of these time-measurements can be pushed to almost any extent. In order to demonstrate this fact, and also to exhibit the high mutual consistency of various determinations, I reproduce another set of records from a different specimen (fig. 84) from which the velocity of transmission is to be determined by the *Differential Method.* In this case a new recorder was taken, with a vibration-frequency of 20 times per second. Hence the distance between any two successive dots represents a time-interval of one-twentieth of a second. The stimulus intensity was again 3. The lowest record gives us the result obtained when the point of application of stimulus was 30 mm. away from the responding pulvinus. The total time T is here seen to be 2·9 seconds. The next record gives us the result when the point of stimulation was at a distance of 20 mm., and the total time T_1 is 1·985

FIG. 84.—Determination of velocity by differential method. Uniform stimuli applied at distances of 30 mm., 20 mm., and directly. Recorder 20 D.V.

second. The third and highest gives the record of direct stimulation, the latent period being shown as ·085 second. Thus

$$V_1 = \frac{30}{2{\cdot}9 - {\cdot}085} = 10{\cdot}7 \text{ mm. per second}$$

$$V_2 = \frac{20}{1{\cdot}985 - {\cdot}085} = 10{\cdot}53 \text{ mm. per second}$$

$$V_3 = \frac{10}{2{\cdot}9 - 1{\cdot}985} = 10{\cdot}9 \text{ mm. per second}$$

It is thus seen that, taking the precautions described, successive determinations of the velocity of transmission may be arrived at which are of great constancy. It may be said of the velocity of transmission in the petiole of *Mimosa* that it is constant with a given specimen, but undergoes some variation with different individuals. It is also subject to modifications induced by season. In winter the velocity is much reduced. The highest velocity which I have obtained with summer specimens of the petiole of *Mimosa* is 30 mm. per second. The lowest, in sluggish specimens, may be as little as 4 mm. per second.

The tabular statement below shows results obtained in twenty-five determinations in different specimens. Maximal stimulus was applied in every case.

TABLE I.—DETERMINATION OF VELOCITY OF TRANSMISSION OF EXCITATION IN VARIOUS SPECIMENS OF *MIMOSA*

No.	Stimulus.	Distance in mm.	True time.	Velocity in mm. per second.
1 ..	3	30	1″	30
2 ..	,,	40	1.4″	28·6
3 ..	,,	20	0.76″	26·3
4 ..	,,	20	0.8″	25
5 ..	,,	40	1.7″	23·6
6 ..	,,	30	1.3″	23
7 ..	,,	30	1.4″	21·5
8 ..	,,	15	0.7″	20
9 ..	,,	30	1.6″	18·8
10 ..	,,	30	1.7″	17·6
11 ..	,,	40	2.4″	16·7
12 ..	,,	15	0.9″	16·6
13 ..	,,	30	1.9″	16·3
14 ..	,,	25	1.6″	15·6
15 ..	,,	40	2.7″	14·8
16 ..	,,	40	2.8″	14·3
17 ..	,,	40	3.0″	13·3
18 ..	,,	30	2.32″	13·0
19 ..	,,	30	2.64″	11·4
20 ..	,,	30	2.96″	10·
21 ..	,,	20	2.1″	9·5
22 ..	,,	20	2.4″	8·3
23 ..	,,	40	5.4″	7·4
24 ..	,,	20	3.3″	6·0
25 ..	,,	10	2.3″	4·3

Having now obtained means for the accurate determination of the velocity of transmission, we next proceed to study the effects of various agencies in inducing changes in the normal rate. And first we must consider the important question of whether or not the velocity is in any degree dependent on the intensity of stimulus. In the corresponding case of conducting animal-nerve there is considerable diversity in the results which have been arrived at. It has been found by some investigators that velocity is independent of the intensity of stimulus. Others have found, on the contrary, that the velocity of transmission increases with the intensity of the stimulus, till with a very high intensity it becomes unmeasurable. The results which I shall here describe will probably throw light on this debatable question. It may be said, in anticipation, that the effects are to some extent modifiable in a definite way by the condition of the conducting-tissue.

If the specimen happens to be in a sluggish condition, then increasing intensity of stimulus will be found to be attended by increasing velocity of transmission. Again, a moderately strong intensity of stimulus is often found to leave, as an after-effect, increased conducting power. That is to say, to a tissue which has been sluggish, stimulation itself imparts a higher conductivity. In these facts there is a remarkable parallelism to what has already been pointed out in the matter of the amplitude of response of the subtonic tissue. In that case we saw that increasing intensity of stimulus gave rise to increased amplitude of response. We saw, further, that following a given stimulus, increased excitability appeared as an after-effect, so that the repetition of an identical stimulus evoked response of enhanced amplitude.

Effect of Increasing Intensity of Stimulus

Returning now to the experimental inquiry into the influence of intensity of stimulus on the velocity of transmission, I reproduce a set of records (fig. 85) obtained with a tissue which was slightly sluggish. The distance of the point of application of stimulus—namely, 20 mm.—was maintained constant, the intensity being varied in the successive experiments. The vibrating recorder had a frequency of 10 per second. The lowest record is the result of a stimulus-intensity of ·5. The total time of transmission is seen to have been 2·1 seconds. The true time is obtained by subtracting from this the latent period, the average value of which is found to be about ·1 second. No appreciable error will be introduced in practice by adopting this average value for the latent period, for its

FIG. 85.—Effect of intensity of stimulus and its after-effect on velocity. Lowest record under stimulus ·5; the next under stimulus 4. Velocity increased under stronger stimulus. Enhancement of conductivity by previous stimulation seen in two upper records under ·5 and 4 respectively. Velocity high and practically the same in both cases.

variations are very slight, being of the order of hundredths of a second. The actual time here taken for transmission is thus 2 seconds, with a stimulus-intensity of ·5.

The record next above gives the result when the stimulus intensity was 4, that is to say, increased to eight times its original value. The total time is now found to be decreased to 1·6 second, the true time after deducting the latent period being thus 1·5 second. The velocity under this increasing intensity is thus enhanced in the proportion of 2 : 1·5, or 33 per cent. To find out if there had been any after-effect of stimulation, a record was once more taken with the original feeble stimulus-intensity of ·5. It will be seen from the third record that the time taken for the transmission of excitation was practically the same as that with the previous strong stimulus, showing that this has made the tissue better conducting and that this property has reached a limit of uniformity. In order to test this conclusion further, a fourth record was taken with the second high stimulus-intensity of 4. It will be observed that the time of transmission is now the same as with the feebler intensity.

From these experiments it will be understood that when the tissue is in a somewhat sluggish or sub-tonic condition the velocity of transmission is enhanced under increasing intensity of stimulation. This, however, reaches a limit under a maximal stimulus the value of which is about 3 units. The following table gives the results of two sets of experiments on the effect of increased intensity of stimulus on velocity :—

TABLE I.—EFFECT OF INTENSITY OF STIMULUS ON VELOCITY IN SUB-TONIC SPECIMENS.

No.	Stimulus.	Velocity.
		mm. per sec.
1	0·5	10
	4	13·3
2	0·5	5·9
	2·5	8·3

The fact that stimulus itself may enhance conductivity in a sub-tonic tissue can be seen in a striking manner in specimens of *Mimosa* which are in sluggish condition. It will there be found that the application of stimulus on the petiole will at first fail to be conducted. If, however, we apply the same stimulus again after the usual interval of, say, 15 minutes, the excitation which failed in the previous case to be conducted will then reach the pulvinus and induce the responsive fall. Stimulus had thus imparted conductivity to the tissue.

Effect of Optimum Condition

We have seen that when the tissue is in a favourable tonic condition the range between minimal and maximal excitation tends to vanish—that is to say, a moderately feeble stimulus induces the same amplitude of response as the maximal. It appears probable that what was found to be true in the case of motile excitability may be equally applicable to conductivity; that velocity of trans-

Fig. 86.—Effect of optimum condition on velocity. Plant raised to temperature 30° C.; records taken under stimulus ·5 (lowest), 2 (highest), ·5 once more (middle). Velocity practically the same.

mission will tend to be constant, even under varying intensity of stimulus, when the tissue is in a favourable tonic condition.

In order to test this induction I next tried the effect of varying intensity of stimulus on a specimen which had been brought to a favourable tonic condition. We have already noticed how the excitability of the plant is enhanced when the surrounding temperature is raised to 30° C. or thereabouts. This was secured by enclosing the experimental plant in a thermal chamber, which was maintained at the uniform temperature of 30° C. The point of stimulation was at a distance of 30 mm. from the pulvinus, and the frequency of the vibrating-recorder employed was 10 times in a second.

In the lowest of the three records shown in fig. 86 a stimulus-intensity of ·5 was employed. The next record, the highest in the figure, was taken with stimulus 2, that is to say, four times the former intensity, and the time of transmission was found to be practically the same as with the feeble stimulus. The third, which is the intervening record, was taken when the stimulus had been restored to its original feeble intensity of ·5. The results demonstrate that these successive experiments, with varying intensities of stimulus, gave practically uniform results in velocity of transmission of 20 mm. per second.

These experiments confirm the conclusion that when the plant is in an optimum condition its velocity of transmission is practically constant, even under varying intensity of stimulus. In ordinary circumstances the velocity increases with increasing intensity of stimulus, till a limit is reached under maximal stimulation.

Disturbing Action of Leakage of Current

On employing the very strong stimulus-intensity of 15 or 20 units, I have sometimes observed a sudden enhancement of the normal velocity. In fact, the time elapsing in these circumstances between stimulus and response was tantamount to the duration of the latent period, the velocity of transmission being thus practically infinite. One thing that was noticeable in such experiments was that instead of gradual and continuous enhancement of the velocity, with increasing stimulus, the enhancement which occurred was sudden and abrupt at a certain high intensity. This justifies the conclusion that, in such a case, the stimulation becomes virtually direct by leakage of currents of relatively high tension.

Effect of Fatigue

It should be stated here that very strong stimulation has a tendency to induce fatigue, the result of which is seen in the reduction of the rate of transmission in subsequent experiments. The effect of fatigue can also be shown under moderate stimulation by reducing the period allowed for rest between two stimulations. In summer the period of complete recovery of conductivity is about 15 minutes. In winter the same process requires from 20 to 25 minutes. In fig. 87 a pair of records is given showing the reduction of the velocity under fatigue. The upper of the two records

Fig. 87.—Effect of fatigue. Upper record when plant fresh; lower record when fatigued.

was taken when the plant was fresh, the distance of the point of stimulation was 10 mm., and the intensity of stimulus was 2. The next and lower of the two records was taken after allowing the incomplete resting-interval of only 10 minutes. It will be seen that fatigue has here prolonged the time taken for transmission of excitation. This prolongation is due chiefly to fatigue of conductivity and partly to the prolongation of the latent period. The variation of the latter factor, however, is relatively insignificant, being only, as already stated, of the order of hundredths of a second. Taking the approximate value of the latent period to be ·1 second, the period required for transmission of excitation through a distance of 10 mm. was here 1·32 second when the plant was fresh; when the specimen was fatigued the period of transmission through the same distance was prolonged to 1·7 second.

TABLE II.—EFFECT OF FATIGUE ON VELOCITY OF TRANSMISSION.

No.	V. in fresh specimen.	V. under fatigue.
1	7·6 mm. per sec.	5·9 mm. per sec.
2	30 mm. per sec.	19 mm. per sec.

The effect of fatigue is also here depicted in an interesting manner in the records of the responsive movements themselves. In the upper record of the fresh specimen the movement is seen to have been vigorous, by the comparative erectness of the curve and the distance between the successive dots, representing the amount of the excitatory fall during periods of one-tenth of a second. The lower curve offers a marked contrast in both these respects.

EFFECT OF TEMPERATURE

If the phenomenon of transmission in the plant is one of protoplasmic change, then any factor that causes physiological variation must have a corresponding influence on its velocity. One such cause of physiological variation is found in change of temperature. If, on the other hand, the propagation had been merely of a hydrostatic blow, then a change of temperature would not have had any marked effect upon it. Thus the accurate determination of the influence of temperature upon velocity of transmission becomes an important consideration in discriminating between the excitatory or mechanical nature of the transmitted change.

I will now proceed to give quantitative results as to

the variation of conductivity, under variation of temperature. The plant was maintained at the required temperatures in the thermal chamber, either by the cooling device or by the electrical appliances for heating, which have been previously described. In fig. 88 time-records are given of the transmission of excitation at the three temperatures of 22° C., 28° C., and 31° C. This experiment was carried out in the Calcutta winter, when the temperature of the room was 22° C. The normal velocity of transmission in the plant was thus, owing to the season, somewhat low. Stimulus of maximal intensity 2 was applied, at a distance of 10 mm. from the responding point. The lowest of the

FIG. 88.—Effect of temperature in enhancing velocity of transmission. Three records, from below upwards, are for temperatures 22° C., 28° C., and 31° C. respectively.

three records gives us the period of transmission at the temperature of 22° C. The next record was taken at 28° C., and the third or topmost at 31° C. It is quite evident from these figures that the velocity is continuously increased under rising temperature. The period taken at 22° C. was 2·94 seconds; at 28° C., 1·69 second; and at 31° C., 1·2 second. We noted in a former experiment that the latent period undergoes a variation with changes of temperature. Thus in a given experiment, while the latent period at 23° C. was ·165 second, at 28° C. it was ·12 second, and at 33° C., ·065 second. These variations are very slight as compared with the total time required for the transmission of excitation. Hence if, in the results of the last experiment, we make the small corrections representing the variations of the latent period, the velocity of transmission at 22° C. will be found to be 3·6 mm. per second, at 28° C. it is increased to 6·3 mm. per second; and at 31° C. it has become 9 mm. per second. Thus at 31° C. it is two and a half times as great as the velocity at 22° C.

The results of this and a few of many other experiments on the influence of temperature on velocity are given in the following table. It need only be said that the effect of rising temperature was always to induce an increase in the velocity of transmission.

EFFECT OF TEMPERATURE ON VELOCITY OF TRANSMISSION.

Number.	Temperature.	Distance.	True time.	Velocity.
1 (Winter specimen)	22° C. 28° C. 31° C.	10 mm.	2·93″ 1·59″ 1·1″	3·6 mm. per sec. 6·3 ,, ,, 9·0 ,, ,,
2 ..	23° C. 33° C.	10 mm.	1·5″ ·8″	6·6 mm. per sec. 12·5 ,, ,,
3 ..	29·5° C. 30·5° C.	20 mm.	2·2″ 1·7″	9·1 mm. per sec. 11·8 ,, ,,
4 ..	27° C. 30° C.	10 mm.	1·0″ ·7″	10·0 mm. per sec. 14·3 ,, ,,
5 ..	29° C. 32° C. 35° C.	20 mm.	2·0″ 1·8″ 1·5″	10·0 mm. per sec. 11·1 ,, ,, 13·3 ,, ,,

VELOCITY OF TRANSMISSION IN *BIOPHYTUM* AND *AVERRHOA*

Employing the method that has been described I have determined the velocity of transmission of impulse in the petiole of *Biophytum*, the average value of which is about 2 mm. per second. Fig. 89 shows a record obtained with a typical specimen. Stimulus was applied at a distance of 50 mm. from the responding leaflet. The record was taken by means of the Oscillating Recorder, the successive dots being at an interval of a second. It will be seen that the response took place 24·5 seconds after the application of stimulus. Making allowance for the latent period, the average value of which in *Biophytum* is

FIG. 89.—Record giving transmission-time in *Biophytum*: Successive dots at intervals of a second.

·4 second, the velocity of transmission in this particular case is 2·1 mm. per second. The velocity in the petiole of *Averrhoa carambola* varies from ·5 to 1 mm. per second.

DIRECTION OF PREFERENTIAL CONDUCTION

In connection with the determination of velocity of transmission of excitation in *Biophytum*, I made the discovery of the curious phenomenon of preferential conductivity. It was found that the state of excitation travelled through the conducting petiole of *Biophytum* with greater facility in one direction than in the opposite. The experiment was carried out by employing a leaflet, situated midway in the petiole, as the motile indicator. Equal stimuli were alternately applied at equal distances to the right and to the left of the leaflet. In one case the excitatory wave was transmitted in a centripetal direction, that is to say,

TABLE SHOWING DIFFERENCE OF VELOCITY IN TWO DIRECTIONS.

Direction.	Specimen I.	Specimen II.	Specimen III.
Centripetal velocity	2·5 mm. per sec.	2 mm. per sec.	1·84 mm. per sec.
Centrifugal velocity	4 mm. per sec.	2·9 mm. per sec.	2·2 mm. per sec.

towards the stem or main axis of the plant. In the other case, excitation travelled outwards towards the tip of the leaf or in a centrifugal direction. From a large number of experiments carried out in this manner it was found that the velocity is greater in the centrifugal direction.

SUMMARY

By applying stimuli of constant intensity, and by allowing proper intervals of rest, successive values of velocity of transmission of excitation are obtained which are constant.

Consistent results are also obtained by the employment of *Differential Method* for the determination of velocity of transmission. The automatic records afford measurement of time as short as ·05 second.

The highest velocity of transmission of excitation that has been found in the petiole of *Mimosa* is 30 mm. per second.

In a sub-tonic tissue the velocity of transmission of excitation is enhanced under increased intensity of stimulus. The tissue becomes a better conductor of excitation in consequence of previous stimulation.

In a tissue in optimum condition the velocity of transmission is the same under feeble and strong stimulation.

Fatigue depresses the rate of conduction of excitation.

Velocity of transmission of excitation is enhanced at a higher temperature.

Excitation is transmitted in both directions; but the velocity is not necessarily the same in the two cases. In *Biophytum* the velocity in the centrifugal direction is greater than in the centripetal.

CHAPTER XII

EXCITATORY CHARACTER OF TRANSMITTED IMPULSE IN PLANTS

The hydro-mechanical theory—Inconclusive character of the anæsthetic experiment of Pfeffer and scalding experiment of Haberlandt—Kühne's experiment showing transmission of excitation under intense stimulation in a rigored nerve—Error introduced by employment of excessive intensities of stimulus—Discriminative polar effect of current in excitation—Block of transmission of excitation by local application of cold—Restoration of normal conductivity by tetanising shock in tissue paralysed by cold—Electrotonic arrest of excitatory impulse—Action of various poisons in inducing block of conduction.

In the previous chapter I referred to the prevailing belief that the transmitted impulse in the plant was hydro-mechanical, unlike the excitatory impulse in the animal nerve. This view has been largely based on two well-known experiments of Pfeffer and Haberlandt. In the former of these, the effect of strong stimulus was found to travel over chloroformed parts of the petiole. Pfeffer assumed that the conductivity of this portion must have been abolished, since chloroform is known to abolish motile excitability. In the experiment of Haberlandt an intervening tissue was killed by scalding; in spite of this, stimulus was found to be transmitted across the scalded area.

From these two experiments it was inferred that the impulse which was transmitted could not have been of a true excitatory nature. It was held that, instead of this, a strong stimulus had given rise to a variation whether of increase or diminution of hydrostatic pressure. This variation of pressure, it was assumed, had been hydro-mechanically transmitted, and on reaching the distant pulvinus had inflicted on it a blow which had proved as effective as if a mechanical stimulus had been applied locally. It is thus held that in *Mimosa* there is a *mere transmission of stimulus but no transmission of excitation.*

I shall, however, be able to show that the two experiments referred to are not as conclusive as has been supposed. But before doing this it is as well to examine the basis of the hydro-mechanical theory. According to the Dutrochet-Pfeffer theory, migration of water is the sole cause of propagation of stimulus, transmission being due to movement of water in the vascular bundles. When a wound is made in the stem causing an incision of a vascular bundle, fluid exudes from the wound on account of which there is a diminution of the hydrostatic equilibrium in the bundle. There may, again, be a propagation of stimulus caused by exciting the pulvinus, when a certain quantity of fluid given out by the excited parenchyma is supposed to pass into the vascular bundle.

According to Haberlandt, flaccidity ensues in the sensitive parenchyma on direct stimulation of an articulation. Owing to the deformation of the cells a pressure is induced in the conducting-tissues which is propagated along them and which, reaching a new pulvinus, stimulates it as if by a blow from without. Haberlandt compares the transmission of pressure in the plant with that in an indiarubber tube filled with water.[1]

But transmission through long, and more or less closed, capillary tubes is not the same as that through uninterrupted

[1] 'It is still harder to explain the mechanism by which a stimulus is propagated from the relaxed parenchyma of the curving pulvinus to the excitable parenchyma of an adjacent joint, after a single mechanical stimulus or with chemical or thermic excitation. . . And when Haberlandt compares the resulting movement of the sap "to that within an india-rubber tube containing water at a given hydrostatic pressure in which increase of pressure at any point is propagated in the form of an undulatory wave from one end to the other," the anatomical relations of the conducting-cells hardly seem to justify such a presumption. The experiments on the conductivity of *Mimosa* would have to be scrupulously repeated before forming any final judgment.'—Biedermann: *Electro-physiology*, vol. ii. p. 16 (Macmillan).

indiarubber tube having a large bore. In the former case a considerable mechanical disturbance would be necessary to start the hydrostatic wave which can effectively reach a distant point. Haberlandt supposes such a mechanical disturbance to be brought about by deformation of the mass of parenchyma in the stimulated pulvinus or by injury of the stem or petiole. But it is not at all necessary to initiate the excitatory impulse in *Mimosa* by stimulating the pulvinus; such an impulse may be originated in the thin petiole where there is no turgid mass of parenchyma to be deformed. Excitation may be caused, moreover, by the agency of a physiological stimulus which does not cause any injury or give rise to any mechanical disturbance.

Again, as regards the question as to whether the transmitted variation of pressure would always form an efficient cause of excitation, it was found by Macdougal that sudden artificial variation, whether by increase or diminution of hydrostatic pressure, brought about no responsive fall of the leaf of *Mimosa*.[1]

We now return to the detailed consideration of Pfeffer's experiment on anæsthetics and Haberlandt's on scalding. As regards the former it has been assumed that the conducting-power was arrested under chloroform. It has, however, been pointed out by Vines that though a narcotised pulvinus certainly loses its motile excitability, it does not necessarily follow that its conductivity likewise is completely abolished. In fact, instances are known to physiologists in which a tissue whose excitability has been abolished will still persist in maintaining its conducting-power. This circumstance may be demonstrated in the case of plants by taking a specimen of *Biophytum* and applying a strong stimulus to an old leaf the motility of whose leaflets has been abolished on account of age. Though its own leaflets do not afford any motile indication, the excitation is found *conducted* through the petiole of the old leaf, inducing the fall of the leaflets in a neighbouring young leaf.

[1] Pfeffer: *Physiology of Plants*, vol. iii. p. 95 (Clarendon Press).

It is also extremely doubtful whether in the particular experiment with *Mimosa* the conducting-tissue in the interior could have been effectively narcotised by the external application of the anæsthetic. The task would almost be as difficult as narcotising a nerve-trunk lying between muscles, by the application of chloroform on the skin outside! In the case of the plant it is conceivable that after a very long application a small quantity of narcotic may, by absorption, get access to the internal conducting-tissue; but narcotisation in these circumstances can only be partial. In such a case the transmitted effect of a feeble or a moderate stimulus will alone be arrested; but the block will fail to arrest the transmitted effect of intense stimulation. These considerations will probably explain Pfeffer's observation that, while the effect of strong injury stimulus was always transmitted across the narcotised area, a moderate mechanical stimulus was but occasionally transmitted.

In Haberlandt's experiment the conducting-tissue was supposed to have been killed by scalding. If this had really been the case, then it may be supposed that under an exceptionally strong stimulus a hydrostatic disturbance had been transmitted through the dead tissue and caused stimulation of the distant leaf, as a mechanical blow *de novo*. But excitatory transmission in a plant is usually accomplished by a stimulus which is feeble. Strong doubt may also be entertained as to whether the tissue had really been killed. In my own experience I find it extremely difficult to be sure of killing the interior of a tissue by scalding the outside. This derives additional support from certain experiments of Kühne on conduction of excitation in a nerve, the specimen employed being the sartorius of a frog.

'The delicate nerve which enters the middle of the sartorius by one side, divides within the muscle so that the single fibres that constitute the bifurcation branch many times dichotomously. When Kühne threw the broad upper end of the muscle into heat rigor by dipping it into warm oil, the half which remained normal twitched *on cutting the rigored portion with scissors, showing that excitable nerve-fibres could still be mechanically excited between the rigored and dead muscle-fibres,* and thus carry the excitation centripetally into branches which divide above the rigored portion of the muscle.'[1]

In this experiment we have an instance of transmission of excitation through heat-rigored animal tissue parallel to Haberlandt's experiment on transmission through scalded plant-tissue. In both these cases it is evident that the scalded tissues, though under heat rigor, were not really killed; and that the induced block or abolition of conductivity (caused by heat rigor, electrotonus, and so on) is after all relative. There may thus be an effective physiological block for normal intensities of stimulation, which would, however, fail under abnormal intensities of stimulus such as that of a burn or of a cut. In Kühne's experiment the intense excitation of scissors-cut failed to be arrested, though the conductivity of the nerve had been depressed under heat-rigor. Similar considerations will explain how the intense excitation caused by a burn or a cut may be transmitted through the narcotised or scalded areas in *Mimosa*.

The experiment of Kühne shows further that the conductivity may persist even after the abolition of the motile excitability. The rigored muscle is seen to have lost its motility, though the embedded nerve retained a certain amount of conductivity for excessively strong stimulus of a cut.

In turning our attention to Kühne's experiment we realise the error involved in ignoring the factor of intensity of stimulus in the matter of the effectiveness of a given block to the transmission of excitation. The necessity of discarding crude and drastic methods of excitation in researches on variation of conductivity will now have become obvious. The object of our inquiry is not to find whether a violent

[1] Biedermann: *Electro-physiology*, vol. ii. p. 57 (Macmillan).

mechanical disturbance is transmitted to a distance, but the determination of propagation of physiological change, under normal modes of stimulation. By employing stimulus of graduated intensity, it should be easy to determine the character of a given impulse by observing the effects of various physiological depressors in modifying the power of conduction.

In order to bring the question—whether in a plant there is true transmission of excitation or mere passage of a mechanical disturbance—to a satisfactory issue, it is clear that we ought to proceed in the following way: First, we have to inquire whether it is not possible to find modes of excitation for the plant which would be purely physiological, and in which there can be no element of physical disturbance. Transmitted effect in such a case could only be due to propagation of excitatory protoplasmic change. We will next subject the question to the final test of the physiological block which would arrest an excitatory impulse, but could have no effect on the passage of a hydro-mechanical disturbance.

I shall now describe four different lines of investigations each of which furnishes an independent proof of the excitatory character of the transmitted impulse :—

(1) On methods of excitation by the discriminative polar action of electric currents.

(2) On the block of transmission of excitation by local application of cold.

(3) On the electrotonic arrest of excitatory impulse.

(4) On the action of poison in inducing block of conduction.

Characteristic Polar Excitation by Constant Current

If in a muscle-and-nerve preparation of frog two electrodes are applied on the conducting-nerve, at a certain distance from the responding muscle, it is found that on sending a current through the included portion of the nerve excitation is induced, which on reaching the responding muscle brings about contraction. There is no excitatory action in the case when the current is established very gradually. The electrical current, as such, has generally speaking little or no excitatory action. It is only at the moment of its sudden initiation, or sudden cessation, that the excitatory effect is most conspicuously induced. It is found, moreover, that an excitatory effect is induced by the 'make' of the current at the kathode, or the point where the current leaves the nerve; at the 'break' of the current, on the other hand, excitation is induced at the anode or the point of entry.

Precisely parallel effects I find to take place when an electrical current is sent through a portion of the conducting petiole. A detailed description of the polar effects of currents will be given in a subsequent chapter. I shall here only give what is essential to my present purpose.

Two non-polarisable electrodes are applied, the proximal on the conducting petiole at a distance of 10 mm. from the responding pulvinus, and the distal on the parenchyma of one of the leaflets, such tissue being non-conducting. If the current now be gradually applied by continuously increasing the E.M.F. from zero to 3 volts by means of a suitable potential slide, we shall find that there is no excitatory effect. But if the E.M.F. of 3 volts be applied suddenly, the direction of the current being such that the proximal electrode is kathode, we shall find that the kathode now becomes the seat of excitation and the leaf undergoes a responsive fall after the short and definite period required for the transmission of excitation through the intervening distance.

If the experiment be then repeated with the proximal electrode as anode, and the distal indifferent parenchyma as the kathode, we shall observe no excitatory effect. This is because the effective proximal electrode, which is anode, does not excite at 'make.' The excitatory effect will however be found to take place at the anode, but only at the 'break' of current.

Reverting to the hydro-mechanical theory, we are confronted with great difficulties in accounting for the excitatory effects in the petiole initiated locally at the electrodes.

There is no turgid mass of parenchyma here which by its deformation might cause mechanical disturbance. It may be thought that in some way exudation of sap might cause the necessary hydro-mechanical disturbance. The question now arises: How did excretion occur at the kathode at 'make'? As the excitation takes place at the anode at 'break' are we to suppose that exudation also takes place at the anode?

In the often cited instances of hydraulic transmission of stimulus of a violent blow or a cut, mechanical disturbance is necessarily present. But transmission of excitatory impulse is found to take place under polar excitation, in the absence of all such disturbing factors. This will be realised when in Chapter XVII it is shown that in *Biophytum* excitatory impulse is transmitted by the action of an electric current which is so feeble as not to be perceived by the very sensitive human tongue. This would indicate that the effect transmitted here is physiological rather than physical. In a nerve the protoplasmic change which is the basis of excitation is initiated locally at the point of kathode at 'make' and at the anode at 'break.' The protoplasmic change is then propagated from point to point giving rise to the excitatory impulse. In the petiole also, excitatory protoplasmic change is initiated locally at the kathode at 'make' and at the anode at 'break.' And every circumstance indicates a point-to-point propagation of excitatory protoplasmic change in the conducting petiole.

Block of Conduction by the Action of Cold

It has been shown that the velocity of transmission is enhanced by favourable physiological changes due to warmth. Conversely the conductivity is depressed by lowering of temperature, and this depression may become so great as to induce an actual arrest of conduction. The object of the investigation being the influence of cold on conductivity, special care has to be taken that the lowering of temperature does not in any way affect either the excitability of the point of application of stimulus or the motile sensibility of the responding pulvinus. For this reason cold is applied locally on the petiole, half-way between the point of application of stimulus of induction-shock and the pulvinus. The experimental plant was highly sensitive on account of the favourable summer season. An intensity of stimulus of ·5 unit applied at a distance of 30 mm. from the pulvinus was found to be effectively transmitted. The intensity of stimulus actually employed was 2 units, which was maximal. A strip of cloth 10 mm. in breadth was wrapped round the

FIG. 90.—Effect of cold in inducing retardation and arrest of transmission: (1) Normal record; (2) Retardation due to slight cooling; (3) Arrest of conduction brought about by intense cold; (4) Record of direct stimulation.

petiole midway between the point of stimulation and the pulvinus. This was for the purpose of local application of cold by means of cooled water or by means of small fragments of ice.

Successive records were then taken at intervals of 20 minutes, which is more than sufficient for complete recovery from previous stimulation. Record 1 in fig. 90 gives the time-interval between the application of stimulus and response under normal conditions. There are 20·5 seconds spaces, each space representing ·1 second. The latent period is ·15 second. The true time of transmission is thus 1·9 second for 30 mm.; the transmission time through 10 mm. is therefore ·63 second. The next record was taken when the intervening length of 10 mm. in the petiole was moderately lowered in temperature by the application of cold water. This cooling should be commenced immediately after the previous responsive fall of the leaf. This not only gives sufficient time for the localised cooling of the petiole but also avoids the excitatory disturbance of the pulvinus caused by sudden application of cold to the petiole. During the localised cooling of the petiole the leaf erects itself and becomes fully sensitive when the time arrives for the application of the next stimulus. Record 2 exhibits the effect of moderate cooling; the transmission period is now prolonged, the difference between the two records being a time-interval of ·8 second. On the assumption that the effect of cooling had remained localised, it is seen that the lowering of temperature had prolonged the period of transmission through the 10 mm. of the petiole from ·63 second to 1·43 second. The conductivity has thus been reduced by more than half.

In record 3 is seen the effect of further lowering of temperature by placing small fragments of ice on the strip of cloth. The excitatory impulse initiated by the maximal stimulus of induction-shock had hitherto been unfailingly transmitted. But under the action of intense cold the impulse is arrested. In order to show that the abolition is not due to the depression of motile excitability of the pulvinus, record 4 is taken of the effect of direct stimulation. An inspection of the record shows that the motile excitability has undergone no change. It is thus clear that the impulse initiated by the stimulus has been arrested by the physiological depression of conductivity induced by cold.

Paralysis of Conductivity and Restoration by Tetanising Shock

In connection with this subject I came across the interesting phenomenon of paralysis of conductivity as an after-effect of intense cooling. After obtaining the record of the block under local application of cold, the fragments of ice were removed and the cooled portion of the petiole allowed to regain the temperature of the room, which must have been accomplished in the course of 20 minutes. After this, on taking the record of the transmitted effect of stimulus I found that the block of conduction was still persistent. The conducting-power of the benumbed tissue is thus paralysed for a period which generally lasts for about 45 minutes. I have, however, discovered the very suggestive fact that the lost conducting-power can be very quickly restored by subjecting the paralysed portion of the petiole to the action of tetanising electric shocks.

The Electrotonic Arrest of Excitatory Impulse.

If in a nerve-and-muscle preparation a constant current be maintained in an intervening tract between the point of stimulation and the responding muscle, this current is found to act as a block to the passage of excitation. With moderate intensity of current the block of conduction is due to the depressing action of the anode. For demonstration of electrotonic block of nerve-conduction, an intensity of stimulus is found which is effective under normal conditions. But during the continuation of the blocking current the excitatory impulse due to the testing stimulus is found to be arrested. The transmission is, however, renewed on the stoppage of the current. It is of special interest to have thus at our disposal a physiological block which can be put 'on' and 'off' at will and many times in succession. The alternate transmission and its arrest then affords a very striking demonstration of the excitatory character of the propagated effect.

I have found a similar block of conduction induced in the plant by electrotonus. Various forms of testing stimulus may be employed. It is, however, more satisfactory to employ a form of stimulus the intensity of which may be either gradually increased or maintained constant. These requirements are fulfilled by thermic and electric modes of stimulation. I will now describe two typical experiments on electrotonic block, in *Biophytum* under thermal stimulation and in *Mimosa* under electric stimulation.

Biophytum.—In this plant we have a whorl of leaves bearing sensitive leaflets. Stimulus is applied on the stem by means of the electro-thermic stimulator. The intensity of stimulus is so graduated as to cause an excitatory impulse to traverse the petioles, effecting the fall of the leaflets in a centrifugal order. Selecting a leaf, electrotonic block is applied in the middle part of the petiole. When the anode A is to the left, the excitatory impulse is found arrested at A; when the current is reversed, the arrest is found to take place at the new anode A′ to the right. On the stoppage of the blocking current the excitatory impulse is observed to traverse the whole length of the leaf.

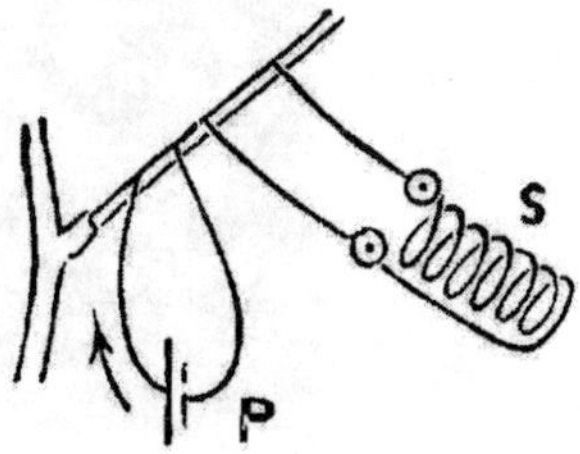

FIG. 91.—Arrangement of the electrotonic block. Polarising circuit P interposed between exciting secondary circuit S and responding pulvinus. The polarising current acts as a block, whether ascending or descending.

Mimosa.—In the next series of experiments a different species of plant and a different testing-stimulus is selected. The employment of an electrical mode of stimulation completely obviates the difficulty of securing a stimulus the intensity of which may be either maintained constant or increased in a known manner. With the help of a sliding induction-coil it is easy to arrive at an intensity of stimulus which is always effective in normal circumstances. The proximal of the two exciting electrodes was placed at a distance of 30 mm. from the primary pulvinus. Half-way between the point of excitation and the pulvinus were placed two polarising electrodes, 5 mm. apart, through which a constant current could be maintained, for the purpose of serving as a block (fig. 91). The first and uppermost of the three records in the next diagram was taken without this block and with a stimulus intensity of 2. It will be seen (fig. 92) that excitation was transmitted as usual, the velocity of transmission in this case being 29 mm. per second. The specimen was exceptionally vigorous and the season the height of summer, which facts account for the high velocity. The blocking current was next introduced. In order to prevent the excitation due to sudden make of

FIG. 92.—Record of effect of electrotonic block. Uppermost record normal; lowest record shows block of transmission of excitation; middle record shows restoration of conductivity on removal of block.

current, the applied E.M.F. was gradually increased from zero to 2 volts, by means of a potential slide. During the passage of this constant current a second stimulus was applied of the same value as before; and it will be seen, from the lowest record, that there was no response, the transmission of excitation being effectively blocked. In order to show that this block would only persist during passage of the current, the latter was next reduced gradually to zero by manipulation of the potential slide. This was carefully done, to avoid the excitation due to sudden cessation of current. On again repeating stimulation, the block was found to be no longer operative (see intermediate record) and response due to transmitted excitation took place as usual.

In the last figure records were taken on a fast-moving

plate. In the next record (fig. 93) a series of response-records to transmitted excitation were taken on a slower-moving plate. The testing stimulus was always the same, the difference being that the electrotonic block was 'off' and

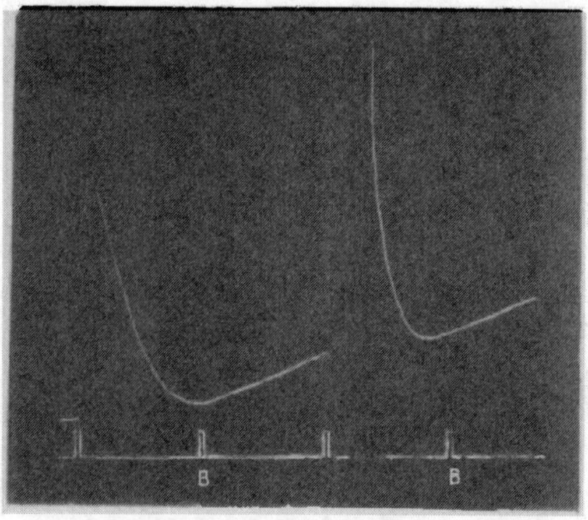

FIG. 93.—Records of transmitted excitation with the block off and on. Arrest of transmitted excitation under electrotonic block at B, B.

'on' alternately. It will be noted that the excitation was invariably arrested whenever the block was applied at B, B.

BLOCK OF CONDUCTION BY THE ACTION OF POISON

(1) *Experiment with Biophytum sensitivum*

Reference has been made to the inconclusive character of Pfeffer's narcotisation experiment. The ineffectiveness of the block, it was explained, might have been due to the thickness of the tissue preventing free access of the anæsthetic to the conducting-elements in the interior. It occurred to me that the physiological block induced by a drug could be rendered more effective in two different ways: first, by the selection of a thin petiole in which access of the solution to the interior by absorption would be less difficult; and, second, by the employment of strong toxic agents like copper sulphate or potassium cyanide solutions. The choice of a strong poison was deemed advisable because the absorption of even a small quantity might in such a case prove effective in inducing depression or abolition of conduction. Application of an anæsthetic, like chloroform, has the drawback that the escaping vapour renders the motile organ insensitive. There is no such disadvantage in the employment of a non-volatile poison like copper sulphate, which by local application would affect the conductivity of the selected portion of petiole without modifying the motile sensibility of the pulvinus. The petiole of *Biophytum* was found to be very suitable for these experiments. Among the whorl of leaves, about half a dozen are found the leaflets of which are fairly sensitive. Graduated stimulus is applied to the stem by means of an electro-thermic stimulator. This latter consists, as explained in a previous chapter, of a V-shaped piece of platinum wire, through which an electric current of suitable intensity could be sent, the circuit being completed for a definite length of time by means of a metronome. The electric current is so adjusted as to give rise to a thermal shock without causing a burn. The effective intensity can be gradually increased by taking advantage of the additive effect of stimulus. Thus in a given case while the thermal shock was singly ineffective, it became minimally effective when repeated four times and maximally effective when repeated eight times. The excitatory impulse originated at the stem, radiated subsequently to the leaves, where the progress of the excitatory waves was visually demonstrated by the serial fall of leaflets in a centrifugal order. It should be mentioned here that the excitation caused by thermal stimulus is most intense ; the block needs to be very perfect to arrest the conduction of such an excitation.

Arrest of conduction by copper sulphate solution.—After determining the value of an effective stimulus, poison was applied to the portion of petiole which is next to the stem. This was done by wrapping a strip of cloth 5 mm. in breadth round the petiole and soaking it with strong copper-sulphate solution. Warm solution was found to be more quickly absorbed than cold. Out of six sensitive leaves, two were subjected to the local action of poison. The selective and local arrest of the excitatory wave in the poisoned leaves would then afford a conclusive demonstration of the physiological character of the transmitted impulse. It should be remembered that the absorption of poison is likely to be a slow process. Hence the physiological block of conductivity induced by poison will become increasingly effective with the duration of application. We may therefore expect the following sequence of events after the application of poison on a narrow zone of the petiole :—

(i) There will be no noticeable variation of conductivity at the beginning.

(ii) After the lapse of a certain length of time the quantity of poisonous solution absorbed will be sufficient to induce a certain depression of conductivity. The minimal excitation which was formerly transmitted will now undergo an arrest. But the blocking action will not be sufficiently great to arrest maximal excitation.

(iii) After a still longer interval, the depression of conductivity induced by poison will be very great. Even a maximal excitation will now be practically arrested.

(iv) This arrest will be due to the abolition of conductivity in the localised poisoned zone. The conductivity of the petiole beyond the poisoned area, and the motile excitability of the leaflets will remain unaffected.

I shall now describe experiments on the effect of poison on conductivity. The experiment was repeated with twelve different specimens of *Biophytum*, all of which gave similar results. The following experiment may be taken as representative of the rest. Among the whorl of leaves in the particular specimen of *Biophytum* there were six which were fairly sensitive. The additive effect of four successive thermal shocks applied on the stem was found to be sufficient to cause excitatory fall of the leaflets in all the six leaves.

Two out of the six leaves were subjected to the local action of poison in the manner previously described :—

(1) The testing stimulus was applied 15 minutes after the application of poison. Excitatory impulse was found to traverse all the leaves. Absorption of the particular poison in the course of 15 minutes was too slight to induce any marked depression of conductivity.

(2) The experiment was repeated after half an hour under the same stimulus as before—namely, the additive effect of four shocks. The excitatory impulse was found transmitted in the four normal, but arrested in the two poisoned leaves. The block in the two leaves was, however, forced by the application of a stronger stimulus due to the additive effect of six shocks.

(3) After two hours the conductivity of the poisoned portion of the petiole was found practically abolished in the two leaves. In these there was no transmitted effect, even under the maximal stimulus of sixteen additive shocks. The untreated leaves exhibited vigorous conduction, the leaflets undergoing their serial fall with great rapidity.

(4) In order to show that the absence of the transmitted effect in the two leaves was due to *local* loss of conductivity of the treated area, and not to the loss of motile sensibility of the leaflets, stimulus was applied on the petiole beyond the poisoned zone. The leaflets exhibited their normal excitatory fall.

Mercuric chloride solution.—After obtaining the block of conduction by the action of copper sulphate, I tried the effect of various other poisons, some of which were found to be very virulent in their action. Mercuric chloride solution, for example, abolished the power of conduction in a much shorter time than copper sulphate. The only drawback in the application of mercuric chloride is that it exerts an excitatory action, the transmitted effect of which induces fall of the sensitive leaflets, which often remain persistently closed.

Solution of potassium cyanide.—There is no such excitatory action induced by this reagent; its toxic power in abolishing conductivity is however very great. A strong solution is applied on a portion of the petiole in the usual manner, the leaflets remaining open and fully sensitive all the time. After an interval of an hour it is found that the effect of a strong thermal stimulus applied in the stem fails to be transmitted across the poisoned zone. The sensibility of the leaflets is however found unaffected. The leaflets in those petioles which have not been poisoned exhibit vigorous response to transmitted excitation.

(2) *Experiment with Mimosa*

I was next desirous of testing the effect of poison on a different species of plant and verifying the result by means of automatic records. The effect of copper sulphate solution in abolishing conductivity was not, as we saw, immediate; it required time to have its toxic effect fully developed. Hence it appeared of much interest to test, by means of successive records, the progressive diminution of conducting power culminating in actual arrest. The specimen employed for these series of investigations was the petiole of *Mimosa*. Successive records of transmission-time were taken at intervals of 20 minutes, before and after subjecting an intermediate portion of the petiole to the action of poison. Effective stimulus of the induction-shock was applied on the petiole, generally at a distance of 30 mm. from the responding pulvinus. The intensity of stimulus employed was maximal, being 2 units. The first record of the series gives the velocity of normal conduction; the second and the subsequent records exhibit progressive action of the toxic agent. This latter was applied on a strip of cloth 10 mm. wide, wrapped round the petiole midway between the point of stimulation and the responding pulvinus.

Copper sulphate solution.—The normal record (1) in fig. 94 shows response to have taken place 27 spaces after the application of the stimulus, the interval between the successive dots being ·1 second. The total time was therefore 2·7 seconds. Subtracting from this the latent period ·15 second, we obtain 2·5 seconds as the actual time for transmission through 30 mm. The time of transmission through 10 mm. is therefore ·83 second.

Record 2 of the series shows the effect of application of copper sulphate solution for 20 minutes on a portion of petiole 10 mm. in breadth. It is noticed that the transmission-time has been prolonged by ten spaces, *i.e.* by 1 second. Assuming that the effect of poison was

FIG. 94.—Effect of copper sulphate solution in the retardation and final arrest of conduction: (1) Normal record; (2) Retardation caused by 20 minutes' application; (3) Arrest caused by application for 40 minutes; (4) Record of direct stimulation.

localised, it is seen that it had during 20 minutes' application delayed transmission through 10 mm. from ·83 second to 1·83 second. By the absorption of a small quantity of poison the conductivity has thus been reduced by more than 50 per cent.

Record 3 of the series was taken after a further period of 20 minutes. The transmitted effect is seen to be completely blocked by the application of copper sulphate for 40 minutes. In order to show that the absence of response was due not to the abolition of motile excitability of pulvinus but to the block of conductivity of the petiole, a fourth record was taken under direct stimulation, which proves that the motile excitability of the leaf had remained unimpaired.

Mercuric chloride solution.—A series of records was next taken exhibiting the effect of mercuric chloride solution. The power of conduction was found arrested after a period of application so short as 10 minutes. The record obtained was very similar to that given in the next figure.

Potassium cyanide solution.—I next took a series of records which exhibited the effect of strong solution of

FIG. 95.—Abolition of conductivity by the action of potassium cyanide: (1) Normal record; (2) Arrest of conduction after application for five minutes; (3) Record showing arrest of impulse due to very strong stimulus; (4) Record of direct stimulation.

potassium cyanide. Record 1, fig. 95, gives the normal transmission period. Record 2 was taken, as stated before, after allowing 20 minutes for recovery. The poisonous solution was however applied 15 minutes after the previous record; hence in the second record we see the effect of application of potassium cyanide for a period of 5 minutes only. The effect of this poison on the conductivity of petiole of *Mimosa* was so great that even with such a short application the transmission of excitation caused by maximal stimulus of 2 units was completely blocked.

Record 3 was taken with the secondary pushed close to the primary, the stimulus intensity being thus raised to 15 units. Even under this intense stimulation the conduction was found to be arrested. Record 4 was obtained under direct stimulation. The response shows that the sensibility of the pulvinus had undergone no change. It is thus clear that the abolition of response to indirect stimulation was solely due to the abolition of conductivity induced by the action of poison.

The arrest of the transmitted impulse in *Mimosa* by the physiological block induced by cold, by electrotonus, and by local application of poison, completely disproves the hydro-mechanical theory. The results of the various investigations that have been described lead on the other hand to the conclusion that the transmission of excitation in the plant is a process fundamentally similar to that which takes place in the animal, being in the one case as in the other a propagation of protoplasmic change.

Summary

Excitatory reaction is initiated in the petiole of various sensitive plants by the discriminative polar action of an electric current. Excitation is induced at the kathodic point at 'make' and at the anodic point at 'break.'

Transmission of such an excitatory impulse takes place in the absence of all mechanical disturbances.

The excitatory nature of the impulse in plants is further demonstrated by the arrest of conduction brought about by various physiological blocks.

Local application of increasing cold retards and finally abolishes the conducting-power.

Conductivity is for a time paralysed as an after-effect of application of cold. The lost conducting-power may, however, be quickly restored by tetanising electric shocks.

The electrotonic block induces an arrest of conduction

during the passage of the blocking current, the conducting-power being restored on the cessation of the current.

Conductivity of a selected portion of the petiole may be abolished by local application of poison. The abolition of conducting-power takes place slowly under the action of copper sulphate and quickly under potassium cyanide.

CHAPTER XIII

THE POSITIVE RESPONSE

Two opposite kinds of responses, negative and positive—Excitatory contraction, negative turgidity variation, fall of leaf, and concomitant negative electric variation—Positive electric response—Positive or erectile mechanical response—Dual impulses under different forms of stimuli—Exhibition of positive and negative impulses by different plants—Conditions for obtaining positive response—Characteristics of positive impulse—Masking and unmasking of positive effect—Laws of Direct and Indirect effects of stimulus.

WHEN the leaf of *Mimosa* is excited, the lower half of the pulvinus undergoes relatively greater contraction; in consequence of this differential action the leaf falls down. There is a concomitant expulsion of water from the excited cells, with diminution of turgor. I have shown elsewhere [1] that the excitatory reaction of plant tissue may also be detected by a method altogether different—namely, by means of electric variation. As regards the sign of electrical change, the excited point is found to become galvanometrically negative; a similar electrical change is also known to take place in the excited animal tissue. The excitatory electromotive change of galvanometric negativity is therefore the same in the plant and the animal.

Excitation in a plant tissue is thus characterised by concomitant effects of contraction, diminution of turgor or negative turgidity variation, mechanical fall of the leaf, and by the electrical response of galvanometric negativity. For the sake of clearness we shall designate this normal excitatory effect as the *negative response*.

In recording electrical responses of plants to indirect

[1] Bose: *Comparative Electro-physiology* (Longmans, Green & Co.),

stimulation I was often surprised to find the occurrence—which appeared at first as an abnormal response—of galvanometric positivity preceding the true excitatory reaction of galvanometric negativity. Thus it appeared as if, in consequence of stimulation, there originated two distinct impulses, positive and negative, which travelled with different velocities. The former travelled faster and, reaching the responding point earlier, induced there the response of galvanometric positivity. The negative wave with its slower velocity of transmission reached the responding-point later and induced the true excitatory effect of galvanometric negativity.

It was also found that the negative effect was much the stronger of the two; in consequence of this, if the two impulses reached the responding-point about the same time, the positive was completely masked by the predominant negative. Hence in order to bring out the positive it was necessary to apply the stimulus at a distance. The slow-moving negative then lagged behind the positive to such an extent as not to mask it.

The following conditions are found to favour the exhibition of the positive effect :—

(1) The stimulus should be applied at a distance from the responding point; the response is then found to be diphasic, positive followed by negative. A distance may again be found where, owing to the enfeeblement of transmitted excitatory effect, the negative impulse fails to reach the responding-point; in such a case there is an exhibition of only the positive effect. Conversely, if the stimulus be applied too near the responding-point, the negative effect alone is exhibited, the positive being masked by the predominant negative.

(2) From what has been said it is easy to understand that, with a very highly conducting tissue, the negative or true excitatory effect will be transmitted with great rapidity; it will therefore mask the positive effect. Hence the positive effect is more easily obtained with plants in

which the velocity of transmission of excitation is relatively slow.

(3) As the velocity of transmission of true excitation is greater under stronger stimulus, a feeble stimulation should be employed for the exhibition of positive effect.

The above are the results obtained from the electric mode of investigation. We have seen that the electric variation of galvanometric negativity corresponds to the excitatory reaction of negative turgidity variation, contraction, and concomitant motile-response by the fall of the leaf. The positive electric response should connote effects the very reverse of these—positive turgidity variation, expansion, and concomitant motile-response of erection of the leaf. Hence if in consequence of stimulation two distinct impulses are sent out with different velocities, we should be able to demonstrate their existence in an altogether different manner—namely, by means of two distinct mechanical responses of erection and of fall of the leaf.

The difficulty that confronts us in this demonstration lies in the relatively small amplitude of the positive response, which is liable to pass unnoticed unless a high magnification be employed. It is easier to demonstrate the existence of the positive impulse by the employment of a magnifying optical lever. A very light mirror is fixed to the fulcrum rod of a lever, one arm of which is attached to the motile leaf or leaflet. Under excitatory fall there is produced a rotation of the fulcrum rod with its attached mirror; a spot of light reflected from the mirror thus exhibits a responsive down-movement, indicative of normal negative response by contraction. An erectile positive response, on the other hand, is recognised by the movement of the spot of light upwards. Responsive movement may in this manner be magnified from a hundred to a thousand times. The optical method is simple and efficient. It is also well suited for purposes of demonstration before a large audience.

Great difficulties are, however, encountered in obtaining

a *record* of the positive response by means of a writing-lever. The weight of the lever stands in the way of obtaining any high magnification, especially in the case of leaflets, where the force of responsive expansion or contraction is very slight. By the employment of an extremely light recording-lever made of aluminium, I was able to secure a magnification of twenty times, which was the highest that could be obtained in the circumstances. With very good specimens this magnification is often sufficient to exhibit the positive effect. In order to reduce friction and obtain time-records, the recording-plate was made to oscillate to and fro once in a second or once in 2 seconds. The successive dots in the record thus indicate intervals of 1 or 2 seconds. It has sometimes been possible to employ the Resonant Recorder with long writing-index for recording the positive response of *Mimosa*. The vibration-frequency of the writer in such a case was five times in a second.

Having thus secured two different methods of observation by means of optical and recording levers, I proceed to demonstrate :—

(1) That whatever may be the form of stimulation employed, two impulses are transmitted, of which the positive travels with a higher velocity than the negative.

(2) That such double impulses are exhibited not by any particular plant but by various species of plants. The phenomenon may therefore be regarded as universal.

Dual Impulses under Different Stimuli

The positive effect, as previously stated, may be separated from the negative by taking advantage of the different rates of transmission of the two impulses. The lag of one impulse behind the other may be increased (1) by taking a specimen in which the velocity of transmission of excitation is low, and (2) by the application of the stimulus at a distance from the responding organ. It should be remembered that the velocity of transmission is not only different in different species of plants, but varies to a certain extent in different individuals of the same species. Again, the transmission under feeble stimulus is slower than under strong stimulus. This accounts for the result frequently obtained—that the propagation-time is much quicker under the strong stimulus of thermal shock than under the moderate stimulation by constant current.

Another interesting fact which has attracted my attention is that the velocity of transmission is, generally

FIG. 96.—Positive followed by negative impulse in *Biophytum*, caused by single indirect thermal stimulus. In this and in the following records, 'down' curve represents positive, and 'up' curve negative, response. Frequency of oscillation once in a second.

speaking, slower in the stem than in the petiole. There also seems to be a loss of time when excitation has to pass from the stem to the leaf. Taking advantage of these facts we may, when desired, obtain long periods of transmission by applying stimulus of moderate intensity on the stem.

I will now describe experiments giving quantitative results, which demonstrate the occurrence of two distinct impulses under different forms of stimuli, the experimental specimen being *Biophytum sensitivum*.

Thermal stimulus.—The electro-thermic stimulator was employed for the application of thermal shocks. The record was taken by means of the Oscillating Recorder, the frequency of oscillation being once in a second. Successive dots are thus at intervals of a second. In the experiment, the record of which is given in fig. 96, stimulus was applied on the petiole at a distance of 50 mm. from the responding leaflet. It is seen from the record that in answer to the stimulus two distinct responses of opposite signs occurred in succession. The positive or erectile response (represented by down curve) is seen to have occurred 1·5 second after the application of the thermal shock; the excitatory or negative response took place much later, that is to say, 22 seconds after the shock. The velocity of the positive impulse is here 33 mm. per second, that of the negative being 2·3 mm. per second.

From the experiment just described it is clear that a single stimulus gives rise to two impulses, positive and negative. The positive travels at a faster rate and gives rise at the responding-point to the erection of the leaflet indicative of positive turgidity variation. The negative or the excitatory impulse travels at a slower rate, inducing at the responding organ negative turgidity variation, contraction, and the fall of leaflet.

It was stated that the transmission-time is relatively long when stimulus is applied on the stem instead of on the petiole. This will be seen clearly from the results of experiments which I shall now describe. Thermal stimulus was first applied on the stem, the distance of the responding leaflet being 10 mm. of stem and 20 mm. of petiole. The positive response took place 3 seconds and the negative 21 seconds after the application of the thermal shock. Stimulus was next applied on the petiole at a distance of 20 mm. Had the velocity of transmission in the stem and the petiole been the same, then the negative impulse would have reached the leaflet after an interval of $\frac{20}{30} \times 21$ seconds or 14 seconds. Instead of this, the transmission-period in the petiole was found to be much shorter—namely,

6 seconds. A statement of the results of two different experiments is shown in Table I.

TABLE I.—SHOWING DIFFERENCE OF TIMES OF TRANSMISSION THROUGH STEM AND PETIOLE

No.	Distance.	Transmission-time for +impulse.	Transmission-time for—impulse.
		seconds	seconds
1	(a) Stem 10 mm. and petiole 20 mm.	3	21
	(b) 20 mm. of petiole only.	2	6
2	(a) Stem 15 mm. and petiole 20 mm.	3	27
	(b) 20 mm. of petiole only	2	5

Chemical stimulus.—The next record was taken under the stimulus caused by the application of a drop of hydrochloric acid on the petiole, at a distance of 30 mm. from the

FIG. 97.—Positive response followed by negative in *Biophytum* by the application of chemical stimulus. Successive dots at intervals of 1 second.

responding leaflet. The record (fig. 97) shows that the positive response occurred 1 second and the negative 14 seconds after the application of the stimulus.

Induction shock.—Stimulus was applied on the petiole at a distance of 30 mm. The positive response took place

2 seconds and the negative 14 seconds after the application of the stimulus.

Constant current.—Stimulation was here induced by the 'make' of kathode. The intervening distance was 20 mm.: the positive impulse traversed this length in 3 seconds, the negative requiring a longer period—namely, 9 seconds.

Condenser discharge.—The distance of the responding leaflet from the point of application of stimulus was 25 mm.: the positive impulse reached the leaflet 4 seconds after the discharge-shock, but the transmission time of the negative was much longer—namely, 15 seconds.

TABLE II.—SHOWING TRANSMISSION OF POSITIVE AND NEGATIVE IMPULSES IN *Biophytum* UNDER DIFFERENT FORMS OF STIMULI

No.	Organ of trans-mission.	Stimulus.	Distance in mm.	Transmission-time for +impulse in seconds.	Transmis-sion-time for —impulse in seconds.
1	Petiole	Thermal	20	2	5
2	,,	,,	20	2	6
3	,,	,,	50	2	29
4	,,	,,	40	1	30
5	,,	,,	65	2·5	50
6	,,	,,	50	1	22
7	,,	Chemical	30	1·5	14
8	,,	,,	60	7	16
9	,,	Induction-shock	30	2	15
10	,,	,,	60	9	23
11	,,	Condenser discharge	25	4	15
12	,,	Constant current	20	3	9
13	Stem and petiole	Thermal	35	3	27
14	,,	,,	35	3	30
15	,,	,,	30	2	45
16	,,	,,	30	3	21

I give above a table showing the effect of various stimuli

on different specimens of *Biophytum.* In some cases the stimulus applied was on the stem, the distance mentioned being the sum of the length of stem and petiole through which the two impulses were transmitted.

It has thus been shown that under the action of various modes of stimulation two distinct impulses are transmitted, of which the positive travels faster than the negative. The specimen employed for these demonstrations was *Biophytum sensitivum.* I next proceed to show that these results are not confined to any particular plant but are universally present.

Exhibition of Positive and Negative Impulses by Different Plants

Averrhoa carambola.—The velocity of transmission of excitation in the petiole of this plant is low, being 1 mm. per second or even less. In the first experiment of the series I applied the stimulus of induction-shock at the moderate distance of 10 mm. from the responding leaflet, and obtained automatic record by means of the Oscillating Recorder. The successive dots here are at intervals of 2 seconds. It will be noticed that a responsive movement of erection took place after an interval of three dots or 6 seconds; the negative response occurred after the much longer interval of 20 seconds (fig. 98).

FIGS. 98, 99.—Positive followed by negative response in *Averrhoa carambola* due to indirect stimulation.

In the next experiment with a different specimen stimulus was applied at a distance twice as great as in the previous case, that is to say, 20 mm. The positive response took place 14 seconds, and the true excitatory effect 48 seconds, after the application of the stimulus (fig. 99).

In an experiment where the chemical mode of stimulation was employed, the distance to be traversed was 40 mm. The positive impulse reached the responding leaflet 19 seconds, and the negative impulse 50 seconds, after the application of the stimulus.

Thermal stimulus was applied in another experiment at a distance of 70 mm. The transmission periods for the positive and negative impulses were 22 seconds and 65 seconds respectively. The negative impulse thus lagged behind the positive by as much as 43 seconds.

FIG. 100.—Records showing positive response in *Mimosa* followed by negative. Stimulus was applied on the petiole 30 mm. from pulvinus. Vibration-frequency 5 per second.

Mimosa pudica.—It was stated that in specimens like the petiole of *Mimosa,* where the velocity of transmission of excitation was high, the positive response was liable to be masked by the predominant negative. It is therefore only on rare occasions that I obtained a positive response by stimulating the petiole of *Mimosa.* This is seen in fig. 100, where stimulus of induction-shock was applied at a point on the petiole 30 mm. from the responding pulvinus. The Resonant Recorder having a long writing-index was employed for obtaining the record. The vibration-frequency of the writer was five times in a second, hence successive dots represent time-intervals of ·2 second. It will be seen that the positive response took place ·6 second and the negative 3·2 seconds after the application of the stimulus.

It was stated that, by applying stimulus at a sufficient

distance on the stem, the transmission period could be made as long as desired. This will be seen in fig. 101, where thermal stimulus was applied on the stem of *Mimosa* at

FIG. 101.—Positive response, preceding the negative in *Mimosa*. Stimulus was applied on the stem. Frequency of oscillation 1 per second.

some distance from the pulvinus. The record was taken on a slowly moving plate, by means of the Oscillating Recorder, the successive dots being made at intervals of a

TABLE III.—PERIODS OF TRANSMISSION OF POSITIVE AND NEGATIVE IMPULSES IN THE PETIOLE OF *Averrhoa* AND STEM OF *Mimosa*

No.	Specimen.	Distance.	Stimulus.	Transmission-period for + impulse	Transmission-period for — impulse.
		mm.		seconds.	seconds.
1	*Averrhoa*	70	Thermal	22	65
2	"	130	"	40	95
3	"	10	Induction-shock	6	20
4	"	20	"	14	48
5	"	35	Chemical	21	50
6	*Mimosa*	5	Induction-shock	·5	12
7	"	10	"	·6	9·4
8	"	20	"	1	10
9	"	60	"	2	29
10	"	35	Chemical	5	17

second. The positive response is here seen to take place 4 seconds, and the negative 41 seconds, after the application of stimulus. The negative impulse in this case lagged as much as 37 seconds behind the positive. In Table III will be seen the transmission-periods of positive and negative impulses in different specimens of *Averrhoa* and *Mimosa.*

Conditions Favourable for the Exhibition of Positive Response

I will now describe experiments which bring out the conditions which are favourable for the manifestation of either the positive or the negative response :—

1. *Positive response is more easily obtained under feeble stimulus.*—This is demonstrated by the following experiment

FIG. 102.—Effect of intensity of stimulus in the induction of positive or negative response. Lowest record under stimulus-intensity of 1, middle record under 5, and the uppermost record under 8 units. Vibration frequency 5 times per second.

on *Mimosa,* where successive stimuli were applied at the same distance on the stem, the intensity being gradually increased in a known manner. The distance of application was 10 mm. The vibration-frequency of the writer was 5 times per second. The first and the lowest record of the series (fig. 102) was taken under the stimulus intensity of 1. It will be seen that under this relatively feeble stimulus a positive or erectile response, indicative of positive turgidity variation, was alone induced ·7 second after the application of stimulus; there was no indication whatsoever of the later occurrence of the negative response. After the usual interval of 20 minutes the next or middle record was taken under the enhanced intensity of stimulus of 5 units. We now observe the appearance of negative following the positive. The positive took place ·6 second and the negative 9·4 seconds after the application of stimulus. Finally, when the stimulus-intensity was raised to 8 units, the positive response took place at the same interval as before, namely ·6 second, but the negative or excitatory response took place earlier than in the last case, that is to say, after an interval of 4·6 seconds, instead of 9·4 seconds.

In another experiment with a different specimen, the positive response took place ·5 second after the application of stimulus of intensity 2; there was no negative response. The point of application of stimulus was kept always

TABLE IV.—EFFECT OF INTENSITY OF STIMULUS ON THE POSITIVE AND NEGATIVE RESPONSES IN *MIMOSA*

Specimen.	Stimulus intensity.	Transmission period for + impulse.	Transmission period for — impulse.
		second	
I ..	1	·7	Negative absent
	5	·6	9·4 seconds
	8	·6	4·6 seconds
II ..	2	·5	Negative absent
	4	·5	12 seconds

the same—namely, 10 mm. When the intensity of stimulus was increased to 4, the negative response was found to occur after the positive. The former took place ·5 second, and the latter 12 seconds, after the application of the stimulus.

From these experiments it is seen that while under feeble stimulus we obtain only the positive response, on increasing the intensity of stimulus the negative or excitatory response makes its appearance in succession to the positive. Another noteworthy fact is that while an increasing intensity of stimulus enhances in a marked manner the velocity of transmission of the negative or excitatory impulse, it has

FIG. 103.—Effect of diminishing the distance of point of application of stimulus. Stimulus applied at a distance of 20 mm. gives rise only to positive response (lower record). Reduction of distance to half gives rise to positive followed by negative response (upper record). Vibration frequency five times in a second.

little or no effect on the velocity of the positive impulse. Thus in the first experiment of the series, while the transmission-period of the negative impulse was shortened from 9·4 seconds to 4·6 seconds by the stimulus increasing from 5 to 8 units, the transmission period of the positive impulse remained unchanged at ·6 second.

2. *Positive response is more easily obtained under a feeble stimulus applied at a distance.*—For the demonstration of this, a feeble stimulus is applied on the stem of *Mimosa* at a distance of 20 mm. from the responding pulvinus. The response is only positive, occurring 1 second after the

application of the stimulus. When the distance of application is reduced to half, we find that the negative or excitatory response makes its appearance in succession to the positive. The positive is seen to take place ·8 second and the negative 10 seconds after the application of the stimulus. It will be noticed that the reduction of the distance to half causes only a slight diminution in the transmission-period of the positive impulse—from 1 second to ·8 second. The transmission-period of the negative impulse, on the other hand, undergoes as we have seen a rapid diminution with diminishing distance.

It has thus been shown that when a feeble or moderate stimulus is applied at a relatively great distance there occurs only the positive response. As the distance is reduced the antagonistic negative response makes its appearance, there being at first a considerable time-interval between the positive and the negative. On reducing the distance still further the interval becomes reduced. The positive will then, as we shall see later, become masked by the predominant negative.

Characteristics of the Positive Impulse

I shall now pass under review the various characteristics which distinguish the positive from the negative impulse.

(1) The positive impulse travels much faster than the negative.

(2) The conditions for the exhibition of the positive impulse are that the stimulus should be moderate or feeble and applied at a distance. A semi-conducting tissue again is more suitable for its exhibition than a highly conducting tissue.

(3) While an increase of intensity of stimulus gives rise to enhanced velocity of negative impulse, it produces little or no change in the velocity of positive impulse. Thus Table IV shows that while raising the intensity of stimulus from 5 to 8 units shortened the transmission-

period of the negative from 9·4 seconds to 4·6 seconds, it induced no variation in the transmission-period of the positive impulse which remained unchanged at ·6 second.

(4) When the distance to be traversed is reduced say to half, the period of transmission of the negative impulse is also reduced to half. But with the positive impulse the diminution is very slight. Thus in fig. 103 we find that with a distance of 20 mm. the period for the propagation of the positive impulse was 1 second; when the distance was reduced to half, the transmission-period was reduced not to half a second but to eight-tenths of a second.

It is difficult from these data to arrive at any definite conclusion as regards the nature of the positive impulse. Is this impulse physical or physiological? If the former, could it be hydro-mechanical? In consequence of a feeble stimulus applied at a distance, we have at the responding pulvinus, a positive turgidity variation, expansion and erection of the leaf, evidently brought about by the forcing in of water. This presupposes a forcing out of water somewhere else, probably at the point of application of stimulus. It may be supposed that an active contraction occurred in the plant-cells under excitation, in consequence of which the sap was forced out giving rise to a hydraulic wave. On this supposition the positive impulse is to be regarded as hydro-mechanical. There is, however, no definite experimental proof in support of this theory. I shall presently describe the experiment I devised to settle the question. Unfortunately the results obtained were not as decisive as I hoped they would be.

It should be clearly understood that even if the positive impulse were found to be hydro-mechanical, it has nothing whatever to do with the propagation of the excitatory or negative impulse. For if the negative response took place in consequence of the impact of the positive impulse, then the negative should take place at a definite interval after the positive, the interval being equal to the latent period of the responding pulvinus.

I have shown that the average latent period of the primary pulvinus of *Mimosa* is ·1 second. In unfavourable circumstances even, I have never found it to exceed ·3 second. The latent period of the leaflet of *Biophytum* or *Averrhoa* is about ·4 second. The latent period of the responding organs of the various sensitive plants mentioned may therefore be taken as less than 1 second. If between the positive and negative response there is any relation of cause and effect, then the negative should occur within a second of the positive. But this is by no means the case. Referring to fig. 101 we find that the positive impulse reached the pulvinus 4 seconds after the application of the stimulus, and instead of the excitatory response taking place within a second, we find that it did not occur till 37 seconds after! I give below a table which shows the very lengthened periods which are often found to elapse between the arrival of the positive impulse and the initiation of the excitatory response :—

TABLE V.—SHOWING DIFFERENCE OF TIME BETWEEN THE ARRIVAL OF POSITIVE IMPULSE AND THE INITIATION OF EXCITATORY RESPONSE

Specimen.	Transmission-period of + impulse.	Transmission-period of −impulse.	Difference of time.
	seconds	seconds	seconds
Mimosa ..	4	41	37
Biophytum ..	2	29	27
" ..	2·5	50	47·5
Averrhoa ..	22	65	43
" ..	40	95	55

The results just described, together with the fact that we may have a positive response without the subsequent negative (cf. lower record fig. 103), show conclusively that the excitatory response is independent of the positive impulse. The physiological character of the negative impulse has, moreover, been fully demonstrated by various crucial experiments described in the last chapter.

For deciding the question as to the physical or physiological character of the positive impulse, I employed the test of local application of cold on the velocity of transmission. It was found that while moderate application of cold delayed the transmission of the negative impulse, it had little effect in retarding the transmission of the positive. Further test by the physiological block induced by excessive cold was, however, less satisfactory. This block not only arrested the negative impulse, but sometimes brought about an apparent abolition of the positive also. This latter result cannot, however, be taken as decisive. The amplitude of positive response is naturally small; under unfavourable circumstances it may have become so diminished as to pass unnoticed. In these circumstances the question of the character of the positive impulse may for the present be left an open one.

Masking and Unmasking of the Positive Effect

It has been shown that as the distance of the point of application of stimulus is reduced, the transmission period of the negative impulse is reduced at a greater rate than that of the positive. When the point of application of stimulus is at some distance, the negative impulse lags considerably behind the positive; but as the distance is reduced, the lag tends to disappear. It thus happens that when the distance of application is sufficiently reduced, the positive effect is masked by the predominant negative.

In an experiment with *Biophytum*, electric stimulus was applied at a distance of 30 mm. from the responding leaflet. The positive response was initiated 2 seconds after the application of the stimulus and persisted for a further period of 12 seconds; the negative impulse then reached the leaflet, giving rise to the reversed negative response (fig. 104). The transmission period of the negative impulse is thus 14 seconds for a distance of 30 mm. The distance was next reduced to half, or 15 mm., and on repeating the stimulus

FIG. 104.—Effect of diminishing the distance of application of stimulus. The lower record shows diphasic response in *Biophytum*, positive followed by negative. By applying stimulus nearer, the positive is masked by the predominant negative (upper record). Frequency of oscillation once in a second.

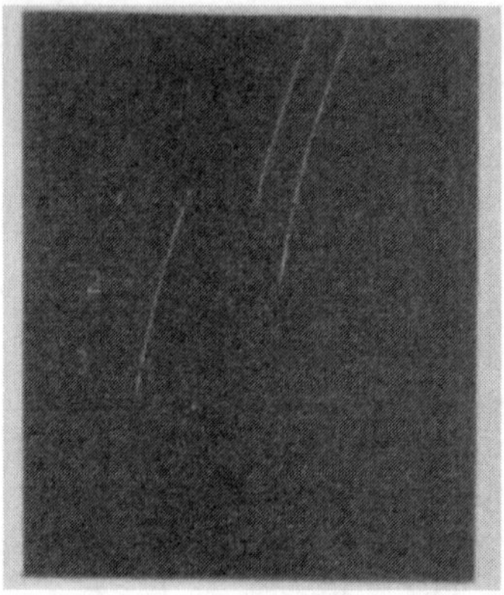

FIG. 105.—Positive (3), diphasic (2), and negative response (1) of *Mimosa* brought about by gradual diminution of distance of application of stimulus.

the positive impulse was found masked, the negative occurring alone after an interval of 6·5 seconds.

I will next describe another experiment carried out with *Mimosa*. Stimulus was applied on the stem at a distance sufficiently great to give only the positive response. This is shown in the lowest record in fig. 105. The point of application of stimulus was then brought somewhat nearer, with the result that a diphasic response took place, positive followed by negative—as depicted in the middle record. And finally, on applying stimulus still nearer, the positive was completely masked by the predominant negative.

We have hitherto dealt with the question of the suppression of the positive by the negative. It is sometimes possible to unmask this suppressed positive by separation of the two impulses brought about by the differential effect of cold. Stimulus is at first applied on the stem at a distance near enough to give rise to the resultant negative response. A strip of cloth is then wound round the intermediate conducting portion of the stem, and the region gradually cooled. The result of this cooling is to reduce the velocity of transmission of true excitation.

I have in this manner often succeeded, by careful and gradual cooling of the intermediate region, in converting a resultant negative into a diphasic—positive followed by negative; further cooling gave rise to the positive alone. In the case of the diphasic response, the result was brought about by the lowering of the velocity of transmission of excitation by cold, in consequence of which the negative lagged behind the positive. In the case of the conversion into positive, the effect was due to the arrest of the excitatory negative impulse.

Direct and Indirect Effects of Stimulus

The effect induced by a given stimulus is modified, as we have seen, by its point of application. A stimulus may be applied directly on the responding organ, or it may be applied at a distance from it. Following the usual terminology we shall designate the former as *Direct Stimulation*, and the latter as *Indirect Stimulation*. When stimulus is applied directly on the responding organ, there is induced an excitatory fall of the leaf concomitant with contraction and negative turgidity variation. This particular reaction we shall designate as the DIRECT EFFECT of stimulus. When, on the other hand, a feeble stimulus is applied at a distance, there occurs only a positive response of erection of leaf with concomitant expansion and positive turgidity variation. For simplicity we shall designate this particular reaction as the INDIRECT EFFECT of stimulus. If the intervening tissue be highly conducting and the stimulus sufficiently strong, then the excitatory negative effect masks the positive. In such a case the effect of indirect stimulation is the same as that caused by direct stimulation. But if the intervening tissue be semi-conducting, or if the stimulus be feeble or applied at too great a distance, then we obtain the positive or INDIRECT EFFECT.

Two diametrically opposite effects may thus be induced by an identical stimulus, depending on direct or indirect application. The existence of the positive or indirect effect of stimulus has hitherto been unsuspected. It must be taken into full account in unravelling the complexities of reaction in a responding organ.

The laws of DIRECT and INDIRECT EFFECTS of stimulus may thus be enunciated :—

The effect at the responding region of a strong excitation transmitted through a short distance, or through a good conducting channel, is negative, being the same as the effect under direct stimulation. The response is by negative turgidity variation, contraction, fall of leaf, and electrical change of galvanometric negativity. This is the Direct Effect of Stimulus.

The effect of feeble stimulus transmitted through a great distance, or through a semi-conducting channel, is positive. The responsive reaction is by positive turgidity variation, expansion, erection of leaf, and electrical change of galvanometric positivity. This is the Indirect Effect of Stimulus.

SUMMARY

A single stimulus gives rise in the plant to two impulses, positive and negative. The positive travels at a faster rate and induces positive or erectile response of the responding leaf. The negative or excitatory impulse travels at a slower rate and induces in the responding leaf a contractile fall. The laws of DIRECT and INDIRECT EFFECTS of stimulus are—

The effect at the responding region of a strong excitation transmitted through a short distance, or through a good conducting channel, is negative; the response is by negative turgidity variation, contraction, fall of leaf, and electric change of galvanometric negativity.

The effect of feeble stimulus transmitted through a great distance or through a semi-conducting channel is positive. The response is by positive turgidity variation, expansion, erection of leaf, and electric change of galvanometric positivity.

CHAPTER XIV

POLAR EFFECTS OF ELECTRICAL CURRENT IN EXCITATION OF PLANTS

Polar excitation in animal tissues—Anomalous reactions in *Protozoa*—Mono-polar method—Current Reverser—Excitatory polar action in plant—Method of record—Effects of ascending and descending currents in animal and plant—Records giving time-relations—Potential slide—Measurement of E.M.F. and current—Potential key-board.

TOWARDS the end of a previous chapter I made brief reference to certain remarkable excitatory effects which I have observed in the plant tissue at the initiation or cessation of a constant current. Electrical currents are well known to have certain characteristic effects in the case of animal tissues. On applying a current of feeble intensity to a muscle-preparation by means of two suitable electrodes, it is found that, at the moment of its sudden starting, an excitatory contraction is initiated at the point of kathode—that is to say, where the current leaves the tissue ; no such effect takes place at the anode, the point at which it enters. With current of feeble intensity there is, again, no excitatory effect either at the anode or the kathode on the cessation or break of current. It should be remembered that these marked exhibitions of excitation are caused by sudden variation of the current. They do not take place if the current variation is made very gradual.

These excitatory effects occur not only when the application of the current is direct but also when it is indirect. That is to say, when the two electrodes are applied on the conducting nerve in a nerve-and-muscle preparation, excitation with a feeble current is initiated only at the kathode at 'make,' this excitation being then conducted along the nerve to cause the contraction of the terminal muscle.

It has been ascertained that this particular type of effect is by no means universal. In working on the polar effects of currents on *Protozoa*, it was found by Kühne, Verworn, and others that the phenomena there displayed are more or less the opposite. Hence it has been assumed by some that the law of polar reaction in unfibrillated protoplasm is entirely different from that which holds good in animal tissues in general.

I shall, however, be able to show that the reactions in the undifferentiated protoplasm of the plant body are identical with those of highly differentiated animal tissues. Experiments with plants, moreover, led me to the discovery that what is known as Pflüger's Law is not a complete statement of the polar reactions that take place in living tissues. For it will be shown that under high electromotive forces a new class of phenomena comes into prominence, culminating in a more or less complete reversal of the normal reactions. From a consideration of these it would appear as if the effects on *Protozoa* were perhaps to be regarded as less anomalous than has hitherto been supposed.

In studying the polar effects of an electrical current on the plant, I shall first take a simple case and employ the pulvinus of *Mimosa* as the contractile organ by which the excitatory action is to be detected. A straight form non-polarising electrode is employed for making the necessary electrical connections. This consists of a glass tube the lower half of which is filled with kaolin paste moistened with normal saline. A wet thread hangs down from the kaolin and makes contact with the plant. The upper half of the tube, above the kaolin paste, is filled with a saturated solution of zinc sulphate; an amalgamated thin rod of zinc dips into the zinc sulphate and serves as the electrode. In the Mono-polar method of experiment one contact is now made at or near the pulvinus, and the other, with a distant indifferent point, lower down on the main stem.

For suddenly starting or stopping a current, or for making an electrode anodic or kathodic, we may use Pohl's Commutator or Reverser (fig. 106). When the commutator is in the middle position, the current is interrupted; when tilted to the left, the current enters the plant through the stem and leaves it by the pulvinus, which thus becomes the kathode. On partially tilting the commutator to the right, the current is broken. On tilting further to the right, the

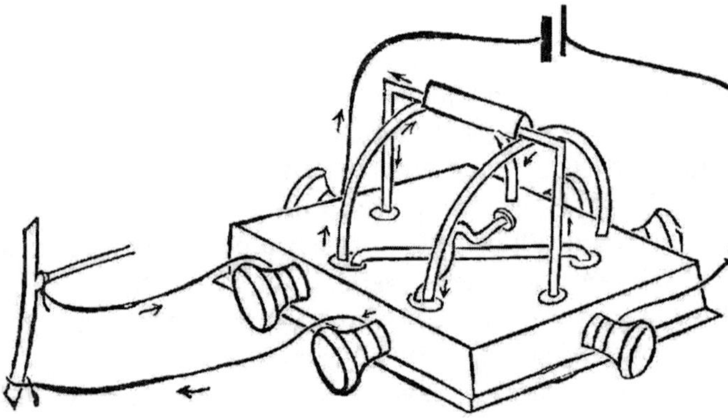

FIG. 106.—Pohl's commutator, for causing make and break of kathode or anode.

current is suddenly reversed and the pulvinus becomes the anode.

For experiment I took a leaf of *Mimosa* which was in a fairly sensitive condition; the electrical connections were made in the manner already described. The electrical resistance offered by the plant tissue between the two points of contact was found to be half a million ohms. An E.M.F. of 4 volts was found effective in inducing excitation at make, when the pulvinus was the kathode. The intensity of the exciting current was feeble, being 8 micro-amperes. A micro-ampere, it should be noted, is one-millionth part of an ampere. When the commutator was turned to the left, the pulvinus was suddenly made the kathode, and this induced an excitatory fall of the leaf. It has been said that the exciting action is in general induced by sudden variation, and not during continuation of current. Hence, if after the excitatory fall the current is continued, the leaf is found to re-erect itself and to have its excitability restored. The current is now broken; this induces no excitation. The commutator is then tilted to the right, the pulvinus being made the anode; this again induces no excitation, nor is there any further excitation at the break of the anode.

It is thus seen that with a *feeble* current the polar effects in *Mimosa* are precisely the same as in animal tissues—namely, excitation only at the make of the kathode.

Having thus described the fundamental reaction of plant tissues under feeble currents, we have still to study the effects of stronger currents and the modifications that may be induced in the normal reactions under changing conditions, both internal and external. Indisputable evidence on such points can only be obtained if it is possible to have automatic records made by the plant itself. This I have been able to secure by the method of record already described, with the addition of a contrivance by which definite signals are marked at the base of the record, at the moment of application of either kathode or anode, its continuation, break, renewal, and so forth. For this a rotating reversing key has a disc attached to the axis. Semi-rotation of this axis to the left or to the right, or its position half-way between, renders the pulvinus kathode or anode, or suffices to break the current. The disc has a thread wound round its circumference, so that when the commutator-axis is rotated in one direction the thread is wound, and in the opposite direction unwound. This thread is ultimately led to, and wound about, a second wheel with a long index, which is to serve as the marker. This second wheel is suitably fixed above the recording-plate, with the tip of the index resting on it at the same vertical line but a little lower than the response-recorder. When the commutator-key is turned to the left, the current is suddenly established, the pulvinus being made kathode. The marker, which has up to this moment been tracing a horizontal line, will simultaneously mark an up-line, as shown in the record. During the continuation of the current, the recorded signal will appear as horizontal but at a higher level. The break of kathode will be indicated by a down-line, stopping at the original level. Anode-make is produced by turning the key to the right, and this is indicated in the record as a down-line passing below the horizontal level. A horizontal line below the general level indicates the continuation of the anode. An up-line reaching the general level indicates the break of anode.

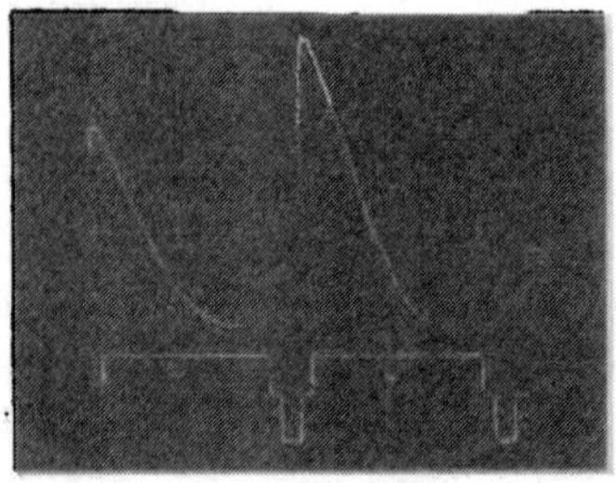

FIG. 107.—Record of polar excitation under feeble current. Response seen to take place only at make of kathode. The line of signal below indicates make or break of kathode or anode. Up-line above normal denotes kathode-make; down-line indicates anode-make.

Two records are reproduced (fig. 107) exhibiting this polar reaction under feeble current in a second and more highly sensitive specimen. The effective E.M.F. was in this case as low as 2 volts. It will be seen that response took place only at kathode-make, there being no effect either at kathode-break or anode-make or break. The next response was taken under the increased E.M.F. of 4 volts. The results depicted are similar to those in the last case, the only difference being the exhibition of an increased excitatory effect at kathode-make due to the higher voltage.

EXCITATION BY ASCENDING AND DESCENDING CURRENTS

In the mono-polar method above described we have the effect of one particular electrode, quite distinct and isolated from the other. But when both the electrodes are placed on a conducting tissue, then there is no such distinctive isolation and the excitatory effect initiated at either of the electrodes may be transmitted to the terminal motile organ, there to induce contractile response. We have an example of this in the nerve-and-muscle preparation of a frog, where under feeble current excitatory contraction of the muscle is seen to take place at make of both ascending and descending currents in the nerve. The terms 'ascending' and 'descending' may be found somewhat confusing, especially in plant-experiments. The meaning will be clear if we remember that the current is said to be descending when it flows towards the responding motile organ. In the case of descending current in a nerve-and-muscle preparation, the kathode is proximal as regards the responding muscle; with the ascending current, the kathode is distal.

As already stated, excitation takes place in nerve-and-muscle preparation at make of both ascending and descending currents. This is explained as due to make of the proximal kathode in the case of descending, and make of the distal kathode in the case of ascending, currents.

It is true that in the latter case the excitation has to traverse the anodic area before it can reach the motile organ. The anode induces, in general, a depression of conductivity which may even block the passage of excitation. In the present case, however, this blocking action is ineffective on account of the feebleness of the current.

The fact that it is the proximal kathode of the descending, and the distal kathode of the ascending, current which form respectively the seats of excitation, may be proved by taking time-measurements of the interval between 'make' of the current and the response of the muscle. With the descending current the interval should be shorter, since in such a case the kathode is proximal. This has been found to be the case.

Characteristics in every way similar are found in experiments with the petiole and pulvinus of *Mimosa*. In this case the petiole contains the conducting tissue and the pulvinus is the indicating motile organ. The parallelism of the two cases will be clearly understood by reference to fig. 108.

With both ascending and descending currents of feeble intensity I have obtained with *Mimosa* excitatory effects at make exactly corresponding to the effects in a nerve-and-muscle preparation. In order to prove that the excitation is really due to make of the proximal kathode in the case of descending, and of distal kathode in that of ascending,

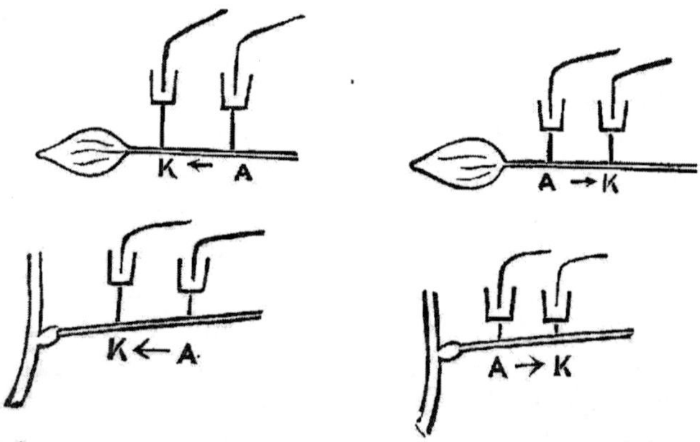

FIG. 108.—Excitation by descending or ascending currents in nerve-muscle and petiole-pulvinus specimens. Motile organs are to the left. Nerve-muscle (upper figure); petiole-pulvinus (lower figure).

currents, I took time-records of these, as seen in fig. 109. The proximal electrode was placed at a distance of 2·5 mm. and the distal at a distance of 15 mm. from the pulvinus. The distance between the two electrodes was therefore 12·5 mm. The time-record was taken in the usual manner, the vibrating recorder employed having a frequency of 20 vibrations per second. The applied E.M.F. was 6 volts. The two records were obtained successively, the upper being the one due to descending, and the lower to the ascending, current. An inspection of the records in fig. 109 at once shows that the response with descending current, the kathode being proximal, took place the earlier of the two.

Theoretically the difference between the two time-records should be equal to the time taken by excitation to travel the intervening distance of 12·5 mm. between the electrodes. This time-interval should be equal to $\frac{12\cdot5}{V}$; where V is the velocity of transmission in the given specimen. The velocity can easily be deduced from the two records themselves. From the upper record we see that response took place after 8 spaces, each representing one-twentieth of a second. The true time taken by stimulus to traverse the distance of 2·5 mm. is therefore $\frac{8}{20}-L$; where L is the

FIG. 109.—Records of responses to ascending and descending make currents in *Mimosa*. In the upper record the current was descending, kathode being proximal to pulvinus. In the lower record the current was ascending, kathode being distal. Frequency of vibrating recorder 20 D.V.

latent period of the pulvinus, the average value of which we found to be ·1 second; hence $V = \frac{2\cdot5}{\cdot3} = 8\cdot3$ mm. per second.

From the second record we find the true time taken by excitation to traverse the distance of 15 mm. to be $1\cdot8 - L = 1\cdot7$ second. $V = \frac{15}{1\cdot7} = 8\cdot8$ mm. per second. The mean velocity obtained from the two records is therefore 8·6 mm. Hence the difference of time in the two cases of excitation by ascending and descending current should be equal to the time taken by the excitation to travel 12·5 mm. the distance between the two electrodes. This difference should therefore be $\frac{12\cdot5}{8\cdot6} = 1\cdot4$ second. That this is really the case can easily be tested by the simple process of counting the time-difference in the number of dots in the two records. On counting we find the difference to be 28 spaces, each representing one-twentieth of a second. That is to say, the time difference is actually 1·4 second, precisely what was inferred from theoretical considerations.

EFFECTS OF ASCENDING AND DESCENDING INDUCTION-SHOCKS

In the case of animal tissues, single induction-shocks of moderate intensity are known, as regards polar action, to be

FIG. 110.—Records showing time-difference between responses due to ascending and descending induction-shocks. Distance between two electrodes 5 mm.; upper record with kathode nearer the pulvinus. Vibrating recorder 20 D.V.

effective at the commencement and not at the termination of the current. Hence excitation here, as in the case of constant current, takes place at the kathodic region. In the case of plants also I find the same to hold good. The exciting electrodes from an induction-coil were placed on two points on the petiole, separated from each other by 5 mm. Two records were taken with single induction-shocks of intensity 2, now in descending and afterwards in ascending directions. In the first or upper of these records (fig. 110) the proximal electrode was the kathode; in the lower, the kathode was distal. It will be seen from these two experiments that if it be the kathode which excited then the arrival of excitation at the pulvinus will be quicker when the kathode is near and later when the kathode is distant. It is clear from the records that this is what obtains ; excitation is earlier when the kathode is proximal and later when the kathode is distal. We find moreover that the difference of time between the moments of arrival of excitation in the two cases is represented by six spaces, each of ·05 second. Hence the delay in the second case is equal to ·3 second, due to transmission through the additional distance of 5 mm. The velocity of transmission is therefore 16 mm. per second, which we have seen is the average velocity of transmission through the petiole of *Mimosa*.

After obtaining these characteristic effects with *Mimosa* under feeble electric currents, we have still to find out whether such effects are peculiar to this plant or whether they are universally present. To answer this question we have to experiment with every species of plant that happens to be provided with a motile indicator. After this has been done we have still to determine the effects of increasing intensity of current brought about by the application of increasing E.M.F.

The Potential Slide

For the application of a graduated increase of E.M.F. up to 12 volts or thereabouts, a Potential Slide is very convenient. This consists of two platinoid wires stretched side by side. Two spring-contacts E and E′ are carried by a slide between the two wires. When a battery of six storage-cells is applied at the terminals of these wires, at A and B, the difference of electric potential at the two points is 12 volts. When the slide is at the extreme end, to the right, the difference of potential between E and E′ will be zero, whereas in being moved to the left the difference of potential or E.M.F. between E and E′ will continuously increase till at the extreme left it is 12 volts, E being, say, positive, and E′ negative. The electrical current enters the plant by E, which is therefore the anode. It leaves by E′, which is therefore kathode. The advantage of the slide is that the E.M.F. applied may be gradually increased from zero to maximum, or decreased from maximum to zero. In certain experiments, where we wish to apply a given E.M.F. without causing excitation, this is essential. For we have seen that a sufficiently graduated increase or decrease of current causes no excitation. If excitation be desired, however, it may be induced by a sudden variation of current brought about by suddenly moving the slide either backwards or

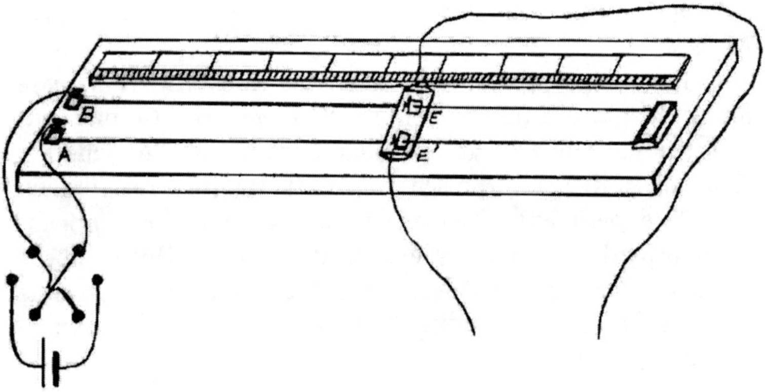

FIG. III.—The Potential Slide.

forwards, or by suddenly completing or interrupting the main circuit (fig. III).

It is important also to know the E.M.F. that has been applied and the intensity of the current flowing through the plant in a given experiment. For the former, a scale fixed on the apparatus may be previously calibrated so as to give the values of the E.M.F. at the different positions of the sliding-contacts, it being understood that the same number of storage-cells are always applied at the terminals. Or we may place a voltmeter to measure the applied E.M.F. between the two sliding-contacts.

For the measurement of the currents that flow through the plant, a micro-amperemeter is used. Each division of this instrument indicates one-millionth of an ampere. The currents which usually have to be measured vary from one to a hundred micro-amperes. A single instrument provided with appropriate shunts enables us to measure all these intensities.

Another point which should be borne in mind is the possibility of induction-currents being generated in the measuring instruments themselves, when the main current is suddenly turned on and off. In order to guard against any disturbance that might be caused in this way, it is necessary at the moment of experimenting on the plant to put the voltmeter and the micro-ammeter out of action for the time. On the completion of each experiment the E.M.F. and the current employed are measured by means of the voltmeter and the micro-ammeter.

In experiments where still higher voltage may be required this potential slide cannot be used, as the wires become unduly heated. We may in such a case modify the potential slide by replacing the platinoid wire with two parallel troughs containing copper-sulphate solution. In place of the spring-contacts two copper plates are here employed, dipping into the solution. These, forming the moving contacts E and E′, are attached to the slide. With this form of electrolytic potential slide a graduated E.M.F. up to 100 volts may be easily obtained. The terminals A and B are connected with a battery of the required number of storage-cells from the installation in the laboratory.

A more convenient arrangement is a portable set of storage-cells, with a Potential Keyboard, by means of which any voltage up to 100 volts may be readily obtained by steps of 2 volts at a time. Small storage-cells of flat form may be obtained 6 cm. broad, 2·5 cm. in depth, and 6 cm. in height. Fifty of these cells, giving a total E.M.F. of 100 volts, may be packed in a small box 25 by 30 cm., having a height of 7·5 cm. Ten of these cells are arranged in units, and forty in four groups of ten each.

As regards the former, the ten cells are arranged in

series, the positive end of the first being connected with the first projecting button. The negative end of each cell 1, 2, 3, and so on, is led to its own button. A key K carries a radial arm, which during a rotation from left to right will make successive connections with the projecting buttons, and thus cause the E.M.F. between E and E′ to rise by increments of 2 volts at a time, the maximum E.M.F. thus obtained by contact with the tenth button being 20 volts (fig. 112). A second key, K′, not shown in the diagram, may in a similar

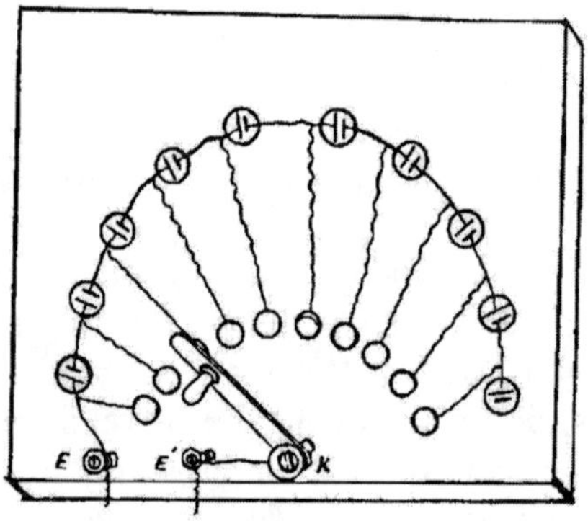

FIG. 112.—The Potential keyboard.

manner make contact with other buttons, which are connected with terminals of cells in groups of ten. Thus the number of cells which may be included between the two electrodes may be varied from 0 to 50, and the derived E.M.F. may thus be raised from 0 to 100 volts by steps of 2 volts at a time.

Having thus found suitable means of applying currents of increasing intensity, we shall in the course of subsequent chapters study in detail the effects of feeble, moderate, and strong currents on various sensitive plants.

SUMMARY

In *Mimosa*, under feeble intensity of current, excitation takes place only at the make of the kathode. In this the polar reaction in plant is the same as in animal tissues.

The effects of feeble ascending and descending currents in the petiole of *Mimosa* are parallel to those in nerve-and-muscle preparations. In both cases excitation takes place earlier when the kathode is nearer the responding organ.

As in animal, so also in *Mimosa*, single induction-shocks of moderate intensity are, as regards polar action, effective at the commencement and not termination of the current. In taking records with single ascending and descending induction-shocks, response is found to take place earlier when the kathode is proximal.

CHAPTER XV

POLAR EFFECTS OF FEEBLE AND MODERATE CURRENTS ON VARIOUS SENSITIVE PLANTS

Polar effects of feeble and moderate currents on (1) leaflets of *Mimosa*, (2) leaflets of *Biophytum*, (3) leaflets of *Neptunia*, (4) leaflets of *Averrhoa carambola*, (5) leaflets of *Averrhoa bilimbi*, and (6) primary leaf of *Mimosa*—Excitation with feeble current only at kathode-make—Excitation at kathode-make and anode-break, under moderate current—Tabular statement of results.

In the previous chapter I described some typical experiments showing the effects of feeble electrical currents on the pulvinus of *Mimosa*. In order to demonstrate that such effects are universal, it will be necessary to extend this investigation to every kind of motile organ possessing requisitive sensitiveness. I here give a list of the specimens employed.

First, the leaflets of *Mimosa* form very sensitive specimens, though there are certain experimental difficulties to be overcome. These leaflets are borne on four sub-petioles that radiate from the end of the main petiole. While the excitatory effect is shown in the primary pulvinus by causing the fall of the leaf, the same effect is seen in the pulvinules by the folding of the leaflets upwards.

Second, the leaflets of *Neptunia* are moderately sensitive. This plant flourishes in tanks, but can with care be grown in pots. The excitatory effect is exhibited by the folding upwards of the leaflets.

Third, the leaflets of *Biophytum sensitivum* are very sensitive and serve the purpose of the investigations in an admirable manner. These leaflets show excitation by folding downwards.

Fourth, the leaflets of *Averrhoa carambola* are only moderately sensitive. They close downwards on excitation.

Fifth, the leaflets of *Averrhoa bilimbi* possess more or less the same kind of excitability as *A. carambola*. Under excitation these leaflets fold downwards.

Sixth, and lastly, we have the primary pulvinus of *Mimosa* whose polar reactions have already been briefly referred to. In this and a subsequent chapter I shall give a more exhaustive series of experiments in connection with the response of the leaf of this plant.

These practically exhaust the list of plant-organs possessed of sensitiveness sufficient for our purpose. It will be my object in this chapter to give a detailed description of the reactions of these various specimens.

We shall find that the character of the reaction will depend on the intensity of the electrical current. With a feeble current a characteristic effect is induced which will be designated as that of Type I. As the acting E.M.F. is gradually increased, the effect is transformed into that of Type II. The characteristic effects of feeble and moderate currents on various sensitive plant-organs will now be dealt with in some detail.

In these experiments, I shall generally speaking employ the Bi-polar method, which possesses many advantages over the Mono-polar method already described. In the Bi-polar method the electrical contacts are made with two points, at or near two different motile-organs. When the commutator is turned to the right, the right contact is suddenly made kathode and the left anode. We are now able to note the excitatory effects, if any, at kathode-make, Km, to the right, and at anode-make, A′m, to the left. The dash affixed in this latter case will always be used to distinguish the contact on the left side. The circuit is now broken, and we observe the effects of kathode-break, Kb, and anode-break, A′b. The commutator is next tilted to the left, making the right anode, and enabling us to note the effects of K′m and Am. The circuit is then broken, producing the results K'b and Ab. It will be noticed that in the second half of this cycle of the operation we subject the experiment to the further test of corroboration by reversal.

The make-effect may be observed simultaneously at both anode and kathode at the beginning of the experiment. The break-effect also may be observed immediately afterwards, provided there has been no excitation during make. But if such excitation has occurred, the current may be maintained till the leaf or leaflet affected has re-erected itself, with accompanying recovery of excitability. The time required for this process is from 10 to 15 minutes in the leaf of *Mimosa* and about 3 minutes in the case of the leaflet of *Biophytum*. It is the sudden variation of the current, and not its continuance, that causes conspicuous excitation. The current may now be interrupted and the break-effect observed.

If it is desired to study the effect of break, apart from a previous over-continued maintenance of current, conceivably inducive of fatigue, we may proceed as follows: By means of the potentiometer-slide already described the current is initiated and brought to a maximum, so gradually as to cause no excitation. The break can now be effected suddenly, in order to note any excitatory effect that may be characteristic. It will be seen that by adopting this method the necessity for maintaining the current, while waiting for the recovery of excitability, is avoided. In order to avoid unnecessary repetition I may state here that I have employed both methods in studying the break-effect, and that as a matter of fact, in ordinary circumstances and within limits, there is but little modification of normal results induced by the previous maintenance of current, the intensity of which at its maximum is of the order of millionths of an ampere.

The distinctive effects of Types I. and II. forming the subject of the present chapter, are brought about the former by relatively feeble and the latter by moderate currents. With a given specimen, increasing the E.M.F. from zero, a minimal value of current is reached at which the effect characteristic of Type I. begins to be evident. On continuing to increase the E.M.F., the same characteristic effect is still for a time obtained, until a critical value of the current is reached above which the effect observed is transformed into that which is characteristic of Type II. There is thus a given range within which we obtain the characteristic effect of a particular type.

The minimally effective current for the induction of any given type of excitatory effect is modified by the condition of the specimen, being relatively low when the plant is highly excitable. Thus age, season, and temperature are all modifying factors. Different species of plants, again, exhibit different susceptibilities to excitation by a constant current. Thus the leaflets of *Mimosa pudica* and *Biophytum sensitivum* are highly susceptible, whereas the leaflets of *Averrhoa* are much less so.

It must also be borne in mind that under the same E.M.F. the intensity of the exciting current will depend on the electrical resistance interposed between the two contacts by the intervening tissue. This resistance, it is obvious, will vary with the distance between the two contacts and the character of the specimen. Thus the resistance offered by a specimen of *Mimosa*, where one contact is at or near the pulvinus and the other on the stem, 7 cm. below, will be of the order of 0·5 million ohm; whereas that interposed by the same length of a thin petiole of *Biophytum* will be something like 15 million ohms. Hence it will be seen that with different plants the value of the E.M.F. applied does not by itself give a correct idea of the intensity of the exciting current. In a series of experiments with the same plant, where the resistance is constant, an increasing E.M.F. does actually connote an increasing current. But in different specimens it should not be assumed that the higher E.M.F. necessarily means a higher value of the exciting current. With a high E.M.F. we may have a feeble current and *vice versa*, on account of the very different resistances offered by different specimens.

To give some idea of the order of magnitude of the current found efficient in causing excitation, I shall, in typical cases, specify its value as read by an interposed micro-ammeter. The unit of current found convenient for these investigations is, as I have stated elsewhere, one-millionth part of an ampere, designated as a micro-ampere or symbolised as 10^{-6} ampere.

Having given a general description of the method employed, we will now study the effects of feeble and moderate currents in inducing excitation. I shall begin by describing the responsive effects seen in the leaflets of *Mimosa*. The bi-polar method has the advantage, as already stated, that the effects at anode and kathode are simultaneously displayed; moreover, the selection of the leaflets instead of the primary pulvinus as indicators of the excitatory effects enables us not only to observe the occurrence of local excitation but also to watch its propagation, as shown in a very striking manner by their serial closure.

LEAFLETS OF *MIMOSA*

Effect of feeble current. The sensibility of leaflets of this plant is considerable, being even greater than that of the leaf. The conducting power of the sub-petiole is of the same order as that of the main petiole. Selecting a young leaf of *Mimosa*, I made appropriate connections with two neighbouring sub-petioles, the length interposed between the two contacts being 4 cm. I shall distinguish the right of these as R and the left as L.

In making repeated experiments on the secondary petioles of *Mimosa* bearing the leaflets, much trouble is occasioned by the travelling of the excitation to the primary pulvinus, in consequence of which the leaf is apt to fall and thereby interrupt the electric circuit. This is obviated by means of a special electrode-holder. A lateral rod, carrying a piece of ebonite, slides up and down on a vertical stand and can be fixed at any height by a screw. The piece of ebonite carries two small plates of brass, each supporting a bored cork through which slides a straight-form non-polarisable electrode. The two pieces of brass can also be adjusted laterally, so as to make the electrical connections with the two sub-petioles at varying distances apart. The main petiole is also supported mechanically and thereby kept

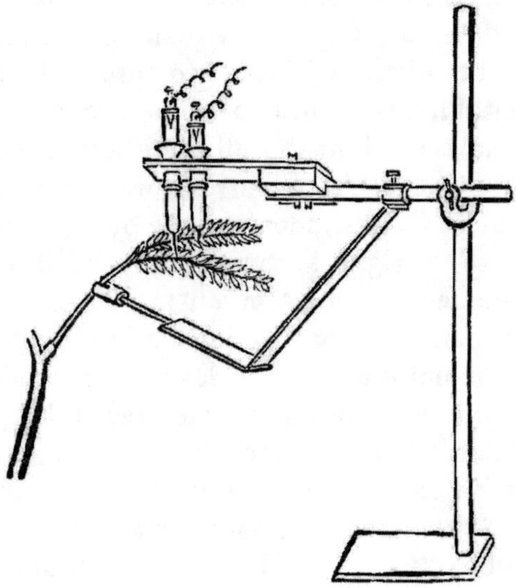

FIG. 113.—The electrode-holder.

from falling, by a strut attachment underneath, as shown in the diagram, which represents two out of four sub-petioles (fig. 113).

Having thus supported the leaf, the next point is to make effective electrical contacts, with middle points on the two neighbouring sub-petioles. In doing this, considerable difficulty is encountered by reason of the waxy coating on the surface of midrib and leaflets. On this account there is apt to be no proper electrical contact between the moist thread of the electrode and the specimen. This obstacle may be overcome, however, by directing a jet of very dilute solution of ether or alcohol—either of which will dissolve wax—

upon the required point and immediately washing it off with water. A minute speck of kaolin paste is then attached to each of the electrodal spots, with the point of a fine brush. The electrical contact with the moistened thread is now found to be perfect.

The E.M.F. is then applied and gradually increased, till the minimally effective intensity is found. If the two sub-petioles be equally excitable, then the same minimal stimulus will excite both to the same extent. If their sensitiveness be slightly different, then the extent of excitatory transmission will be different in the two cases. In the present case the excitabilities of the two sub-petioles were practically the same, and the minimally effective E.M.F. was found to be 4 volts. The exciting current was ·7 micro-ampere. When the current was suddenly made by turning the Pohl's commutator to the right, kathode-make, or Km, was effected on R, and anode-make, or A′m, on L. No effect was found to be induced at A′m, but excitation was induced at Km. This excitation not only caused closure of the pair of leaflets at the kathodic point but was irradiated in both directions from this point. Eleven consecutive pairs of leaflets underwent successive closure in the intra-polar tract, and three pairs in the extra-polar—a fact which may be indicated by the symbol 11 ← → 3. In this convention it will always be understood that ← indicates intra-polar and → extra-polar transmission. Having observed the effects of Km and A′m, the circuit was then broken, thus causing kathode break, Kb, at R, and anode-break, A′b, at L. No excitatory effect was induced.

In order to test these results further, the key was now tilted to the left, making L kathode and R anode. This, it will be noted, is the reverse of the previous arrangement. The excitatory effects were now found to have changed their places, the new kathode K′ at L showing excitation at make, with a similar overflow as before—namely, 11 ← → 7. It is obvious that only by accident could the excitability and conductivity of two different sub-petioles be exactly the same; hence the slight difference in the extent of transmission in the two cases. As regards the sub-petiole R, the anode-make, Am, showed no excitation. On break of the circuit, again, no excitatory effects were observed at either anode or kathode. This cycle of operations was repeated once more, with precisely similar effects.

The results of these experiments will be seen, in proper sequence, in the following tabular statement, which summarises the effects caused by currents acting in a given direction—the direct—and its reverse, for the purpose of corroboration. It should be remembered that at any given moment, when a point on the one sub-petiole is made kathode, K, a point on the other is simultaneously made anode, A′. Excitatory contraction or closure is denoted by C, and the direction of the arrow indicates the transmission in intra- and extra-polar directions. O denotes absence of excitation.

TABLE I.—EFFECT OF FEEBLE CURRENT ON LEAFLETS OF *Mimosa*

Direction of current.		R. Sub-petiole.	L. Sub-petiole.
Direct	Make ..	Km .. C 11 ←→ 3	A′m ....O
	Break ..	Kb .. O	A′b ....O
Reverse	Make ..	Am .. O	K′m....C 11 ←→ 7
	Break ..	Ab .. O	K′b ....O

The results given in the above table may be expressed in a more condensed form as follows :—

TABLE II.—LEAFLETS OF *Mimosa* UNDER FEEBLE CURRENT

E.M.F. = 4 volts; current = ·7 micro-ampere

R. Sub-petiole.				L. Sub-petiole.			
Km	Kb	Am	Ab	K′m	K′b	A′m	A′b
C	O	O	O	C	O	O	O
Excitation formula—**Km**							

In another series the E.M.F. was gradually raised in successive experiments. The specimen in this case was older and less sensitive than the former. The minimally effective E.M.F. was here found to be 8 volts. Successive observations were taken with an E.M.F. of 8 volts, 12 volts, and 14 volts. In all these the excitatory effect only took place at kathode-make. The sole difference was in the increased extent of transmission of excitation with higher voltage. With 8 volts only one pair of leaflets underwent closure; with 12 volts this was increased to 7 pairs; with 14 volts to 9 pairs. With still higher voltage the responsive effects became those of Type II., which will be described presently.

I carried out twenty-five sets of experiments with different specimens of *Mimosa* on the effect of feeble current. The minimally effective E.M.F. was found to vary from 4 to 12 volts, according to the age of the specimen. The young leaflets were excited under a lower E.M.F. than the old. In all these, without a single exception, excitation was found to take place only at the kathode at make.

Hence as the result of these direct and reverse experiments on the effect of feeble current, undertaken with various specimens of leaflets of *Mimosa*, we find one invariable result—namely, that kathode excites at make and that no excitation is occasioned by the break of kathode or by the make or break of anode. The excitation formula of the first type is, to follow the parallel nomenclature in the case of animal tissues, KCC, or, as I have preferred for the sake of convenience, the simpler formula **Km**, which is a short way of saying that excitation took place at the make of kathode. The excitation formula will always be represented by letters in thick type.

Effect of moderate current.—The characteristic effect of the first type is obtained, normally speaking, from an E.M.F. of 4 volts, and continues till the E.M.F. is raised to about 8 volts. In the case of young and excitable specimens we obtain here a transition from the effect of the first to that of the second type. With older specimens, the first type may persist up to 12 volts.

Taking an excitable specimen of leaflets of *Mimosa*, I first applied 4 volts and obtained with it Type I. effect, that is to say, excitation took place only at the make of the kathode. I then raised the E.M.F. to 8 volts. At make, the kathodic point on the right became the seat of excitation, four pairs of leaflets undergoing closure in the extra-polar and six pairs in the intra-polar regions. There was no effect at the anode. At break, however, excitation was initiated at the anode, as evidenced by the serial closure of three pairs of leaflets in the intra-polar and two pairs in the extra-polar regions. This effect was corroborated by a reversal experiment, when the new kathodic point to the left exhibited excitation at make, and the new anodic point to the right exhibited it at break. Thus with current of moderate intensity, excitation took place in the sub-petiole of *Mimosa* at the make of kathode and break of anode. The polar excitation formula of the leaflets of *Mimosa* under moderate current may therefore be expressed as KCC, AOC ; or **Km Ab.**

TABLE III.—LEAFLETS OF *Mimosa* UNDER MODERATE CURRENT
E.M.F. = 8 volts ; current = 1·6 micro-ampere

R. Sub-petiole.				L. Sub-petiole.			
Km C	Kb O	Am O	Ab C	K′m C	K′b O	A′m O	A′b C
Excitation formula—**Km Ab**							

I carried out this experiment on the polar effects of a moderate current on leaflets of *Mimosa* with twenty-five different specimens, and obtained invariably the characteristic effects of excitation only at the make of kathode and the break of anode.

LEAFLETS OF *BIOPHYTUM*

No specimen appears better suited for experiments on polar excitation than the leaflets of *Biophytum*. There is but little difficulty here in securing good electrical connections; the conductivity is moderate, the velocity of transmission being about 3 mm. per second. The susceptibility of *Biophytum* to polar excitation is considerable, an extremely feeble current being sufficient to induce excitation. Its greatest advantage, however, lies in the rapidity of its recovery from excitation, which is practically effected in the course of about 3 minutes. Thus employing the same plant, experiments can be repeated under different conditions without undue waste of time.

Young leaflets are highly excitable and the older specimens less so. Thus we have a wide range of selection according to the particular characteristics which we wish to bring into prominence. When the plant is highly excitable the bi-polar connections are made, on the middle points of opposite leaves, at a distance of 7 cm. from each other. The electrical resistance is high, being in the case described about 15 million ohms. With older and less sensitive specimens the two connections may be made on the same leaf, about 3·5 cm. apart. There will then be about five pairs of leaflets in the intra-polar and about two pairs in each of the extra-polar regions.

When the electrical connections are made with two leaves of fairly equal sensitiveness the excitations induced in the two will be equal. But if one leaf be old and the other young, then a given excitatory effect will occur earlier, or under feebler stimulus, in the younger. The minimally effective E.M.F. will thus be lower in this case than in the older.

Effect of feeble current.—In order to study the effect of feeble current, I took a pair of leaves of *Biophytum* whose sensitiveness was approximately equal. The minimally effective E.M.F. was here found to be 4 volts, the current being of the extremely feeble intensity of ·5 micro-ampere. The excitatory effect was induced at the right contact, when that point was made kathode, six pairs of leaflets undergoing closure in the intra- and two pairs in the extra-polar regions. No effect was induced to the left by anode-make A′m. Nor was there any effect induced at break of current, Kb or A′b. On reversal, the left contact showed the excitatory reaction of K′m, two pairs of leaflets undergoing closure in the intra- and none in the extra-polar regions. The difference in the extent of transmitted excitation as given in Km and K′m is here due to the fact that the left leaf was slightly the older and less sensitive of the two. On break there was no effect induced at either kathode or anode. I carried out in this way more than fifty experiments regarding the excitatory effect of feeble current on the leaflets of *Biophytum*. The minimally effective E.M.F. in these cases varied from 4 to 8 volts, and the range of E.M.F. through which the effects of Type I. persisted was found to be between a minimum of 4 volts and a maximum of 20 volts or thereabouts. In all these experiments of the first type excitation took place only at kathode-make and not at kathode-break, or anode-make or break. This result was without a single exception. The characteristic excitatory formula for *Biophytum*, under feeble current, is thus, as in leaflets of *Mimosa*, **Km**.

The following table gives the result with feeble current :—

TABLE IV.—EFFECT OF FEEBLE CURRENT ON LEAFLETS OF *Biophytum*

E.M.F. = 4 volts ; current = ·5 micro-ampere

R. Contact.				L. Contact.			
Km C	Kb O	Am	Ab O	K′m C	K′b O	A′m O	A′b O
Excitation formula—**Km**							

Effect of moderate current.—Taking another specimen, I applied an E.M.F. of 22 volts between two points in the same leaf, the intensity of current being 2·1 micro-amperes. In carrying out the experiments, I found that the excitation took place at make of kathode and break of anode only. I give a sketch showing these effects of kathode-make and anode-break (fig. 114). The figure at the left shows the depression of leaflets starting from the kathodic point which was on the left. After their recovery the circuit was broken and excitation is seen to have taken place at the anode, on the right, and not at the kathode, on the left. Thus the leaflets of *Biophytum*, with a moderate current,

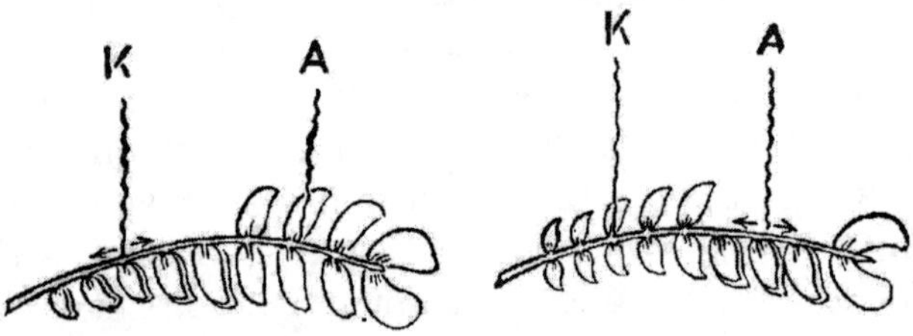

FIG. 114.—Illustration to the left shows excitation induced by the make of kathode, that to the right the effect of break of anode.

gave the same reaction as the leaflets of *Mimosa*, the excitatory formula in both being **Km Ab.** I carried out more than fifty experiments on the effect of moderate currents on the leaflets of *Biophytum*, and the results without a single exception were as has been described.

The tabular statement below shows the results due to a current of moderate intensity on leaflets of *Biophytum* as given in the typical experiment. Each experiment, it should be remembered, was carried through the usual cycle of direct and reverse currents.

TABLE V.—EFFECT OF MODERATE CURRENT ON LEAFLETS OF *Biophytum*
E.M.F. = 22 volts; current = 2·1 micro-amperes

R. Contact.				L. Contact.			
Km	Kb	Am	Ab	K′m	K′b	A′m	A′b
C	O	O	C	C	O	O	C
Excitation formula—**Km Ab**							

Experiments made on leaflets of other plants, in which

the sensibility is much less than that of the leaflets of *Mimosa* and *Biophytum*, will next be described. Amongst these the first plant dealt with is *Neptunia*.

LEAFLETS OF *NEPTUNIA*

This plant is somewhat like *Mimosa pudica* in appearance, except that under natural conditions in Bengal it grows in water, being an inhabitant of the pools amongst the ricefields. In these circumstances it develops curious growths on the stem, which serve as floats. The leaves are much longer than those of *Mimosa*, but considerably less sensitive. There are three pairs of sub-petioles, which are articulated to the main petiole. Electrical connections are made with two points at the middle of two sub-petioles which carry numerous leaflets. The sensitiveness of the leaflets of this plant is much less than in *Mimosa*. The conducting-power is also very moderate, the velocity of transmission of excitation in one of the sub-petioles being 1·5 mm. per second.

Effect of feeble current.—The minimum E.M.F. which will induce excitation is higher in the case of *Neptunia* than in *Mimosa*, being 20 volts in the typical instance now to be described. The resulting current was 7 micro-amperes. On going through the usual cycle of direct and reverse experiments, it was found that the kathode alone excited during make. From experiments with five other specimens I obtained the same result without exception.

TABLE VI.—EFFECT OF FEEBLE CURRENT ON LEAFLETS OF *Neptunia*
E.M.F. = 20 volts; current = 7 micro-amperes

R. Contact.				L. Contact.			
Km C	Kb O	Am O	Ab O	K′m C	K′b O	A′m O	A′b O
Excitation formula—**Km**							

Effect of moderate current.—Continuing to increase the

E.M.F. still further, I found the characteristic effects of Type II. to take place under an E.M.F. of 30 volts, the resulting current being 11 micro-amperes. Excitation thus takes place, with moderate currents, at the make of kathode and the break of anode. This result was borne out without any exception by a series of experiments on five different specimens.

TABLE VII.—EFFECT OF MODERATE CURRENT ON LEAFLETS OF *Neptunia*
E.M.F. = 30 volts; current = 11 micro-amperes

R. Contact.				L. Contact.			
Km	Kb	Am	Ab	K′m	K′b	A′m	A′b
C	O	O	C	C	O	O	C
Excitation formula—**Km Ab**							

It will be seen that, in this case as in others, with feeble current kathode alone excites at make, and with moderate current excitation takes place at make of kathode and break of anode.

LEAFLETS OF *AVERRHOA CARAMBOLA*

These leaflets are as a general rule but slightly sensitive, and require a comparatively high E.M.F. to induce excitation. The bi-polar connections are here made on two points of the same leaf, separated from each other by a distance of 7 cm. The electrical resistance of the petiole is high, being about 5 million ohms when the connections are as described. The velocity of transmission of excitation, also, is relatively feeble, being in a representative case ·5 mm. per second. Under feeble stimulus the excitation travels through a short distance only; still feebler excitation remains more or less localised.

Effect of feeble current.—With young and excitable specimens the minimally effective E.M.F. is 10 volts, the resulting current being 2 micro-amperes. Generally speaking an E.M.F. of 22 volts and a current of 4 micro-amperes are necessary. On starting the current, a pair of leaflets at the kathode underwent an excitatory fall. No effect was induced at the left anodic contact. At break there was no effect at either. On reversal, the left contact became kathode, resulting in the excitatory fall of a pair of leaflets. This local character of excitation was due to feeble conductivity. It is only under much stronger polar excitation that a moderate amount of transmission takes place. At the anode there was on this occasion no effect at make—nor was there any excitation at break of kathode or anode.

This experiment was repeated in the case of ten different specimens, the result being invariably the same.

TABLE VIII.—EFFECT OF FEEBLE CURRENT ON LEAFLETS OF *Averrhoa carambola*

E.M.F. = 22 volts; current = 4 micro-amperes

R. Contact.				L. Contact.			
Km	Kb	Am	Ab	K′m	K′b	A′m	A′b
C	O	O	O	C	O	O	O
Excitation formula—Km							

Effect of moderate current.—In order to obtain the characteristic reactions of Type II., the E.M.F. has generally,

TABLE IX.—EFFECT OF MODERATE CURRENT ON LEAFLETS OF *Averrhoa carambola*

E.M.F. = 50 volts; current = 11 micro-amperes

R. Contact.				L. Contact.			
Km	Kb	Am	Ab	K′m	K′b	A′m	A′b
C	O	O	C	C	O	O	C
Excitation formula—**Km Ab**							

in the case of *Averrhoa carambola*, to be raised to 50 volts or thereabouts. Thus in a given experiment, under the action of 50 volts and with a current intensity of 11 micro-amperes, excitation took place at kathode at make and at anode at break. The experiment was repeated with ten different specimens, the result being always the same.

The excitatory reaction in this plant is thus characteristically the same as in others already described.

Leaflets of *Averrhoa bilimbi*

The excitability of these leaflets is perhaps slightly greater than that of *A. carambola.* I carried out the usual cycle of observations on the effects of feeble and moderate currents, employing in each case five different specimens. The two bi-polar contacts were made in the case of the same leaf 7 cm. apart.

Effect of feeble current.—The excitatory effect of kathode-make characteristic of Type I. was exhibited with an E.M.F. of 10 volts, the current being 6 micro-amperes. The electrical resistance offered by the petiole of this plant was much less than that of *A. carambola.* All the specimens, without exception, gave the effects characteristic of Type I. under feeble current.

Table X.—Effect of Feeble Current on Leaflets of *Averrhoa bilimbi*

E.M.F. = 10 volts; current = 6 micro-amperes

R. Contact.				L. Contact.			
Km C	Kb O	Am O	Ab O	K'm C	K'b O	A'm O	A'b O
Excitation formula—**Km**							

Effect of moderate current.—When the E.M.F. was raised to 28 volts, with a resulting current of 16 micro-amperes, the effects induced were characteristic of Type II. That is to say, excitation was shown at the kathode at make and at the anode at break. Repetition of the experiment on other specimens gave similar results.

TABLE XI.—EFFECT OF MODERATE CURRENT ON LEAFLETS OF *Averrhoa bilimbi*

E.M.F. = 28 volts; current = 16 micro-amperes

R. Contact.				L. Contact.			
Km	Kb	Am	Ab	K'm	K'b	A'm	A'b
C	O	O	C	C	O	O	C
Excitation formula—**Km Ab**							

PRIMARY LEAF OF *MIMOSA*

I have already described in the previous chapter the polar effects of feeble currents on the pulvinus of *Mimosa*. In that case the mono-polar method of experiment was employed. The bi-polar method will now be taken up, enabling us to observe the make-effect of anode and kathode in two different organs simultaneously, and, in the same way, the break-effect; all these to be further corroborated by reversal experiments.

Effect of feeble current.—The bi-polar connections are made either directly with, or in the close vicinity of, two different pulvini, at varying heights on a single main stem of *Mimosa*. With highly excitable specimens the polar effects characteristic of Type I. may here be obtained with as low an E.M.F. as 2 volts. In another case the applied E.M.F. was 4 volts, and the resulting current 4 micro-amperes. On going through the usual experimental cycle—with direct and reverse currents—it was found that the kathode alone exhibited excitation at make. There was no excitation either at kathode-break or anode-make or break. Similar experiments were carried out on some fifteen different plants of *Mimosa*, the results without exception being as described. The only variation seen in the case of different individuals lay in the value of the minimally effective E.M.F., which in less excitable specimens was sometimes as high as 8 volts.

TABLE XII.—EFFECT OF FEEBLE CURRENT ON PULVINUS OF *Mimosa*
E.M.F. = 4 volts; current = 4 micro-amperes

R. Leaf				L. Leaf			
Km C	Kb O	Am O	Ab O	K'm C	K'b O	A'm O	A'b O
Excitation formula—**Km**							

Effect of moderate current.—Under an increasing E.M.F. the characteristic polar effects seen in the pulvinus of *Mimosa* are, as in other cases, transformed into those of

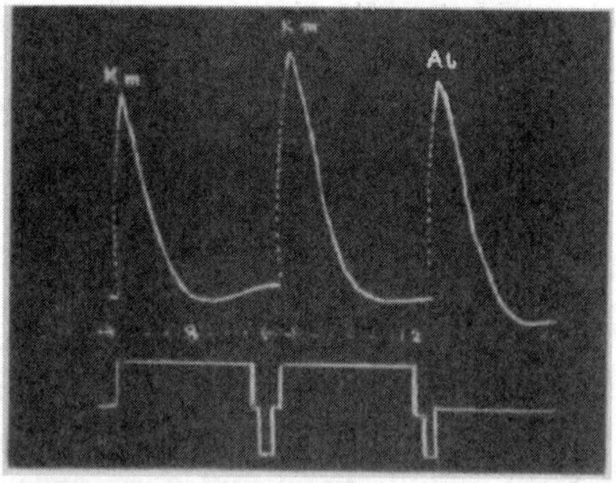

FIG. 115.—Under 8 volts excitation at kathode-make; under 12 volts excitation at kathode-make and anode-break.

Type II. Thus in a particular case the E.M.F. applied was 10 volts, the resulting current being 8 micro-amperes. In carrying the experiment through the usual cycle of direct and reverse currents it was found that the kathode excited at make, and the anode at break. In fig. 115 are shown a series of records given by a different leaf under kathode-make, kathode-break, anode-make, and anode-break. It will be seen that under a feeble current, due to an E.M.F.

of 8 volts, the reactions are those characteristic of Type I. But when the acting current was increased by applying the higher E.M.F. of 12 volts, the responsive characteristics, as was to be expected, changed to those of Type II. Whereas under feeble current the kathode alone had excited at make, now under moderate current excitation occurred at both kathode-make and anode-break.

These distinctive effects of the moderate current were obtained with no fewer than fifteen different specimens in spring, and were thus shown to be characteristic of plants in normal condition. Later in the year, however, seasonal changes were seen to induce certain modifications of the polar effects; these will be fully described in another chapter.

TABLE XIII.—EFFECT OF MODERATE CURRENTS ON *Mimosa*
E.M.F. = 10 volts; current = 8 micro-amperes

R. Leaf.				L. Leaf.			
Km	Kb	Am	Ab	K′m	K′b	A′m	A′b
C	O	O	C	C	O	O	C
Excitation formula—**Km Ab**							

It will now be advisable to recapitulate the results described in the present chapter, in order to show how universal these phenomena are. I will first give a tabular statement of the effect of feeble current :—

TABLE XIV.—EFFECT OF FEEBLE CURRENT ON VARIOUS SPECIMENS OF SENSITIVE PLANTS

Type I.—Excitation formula **Km**

Specimen.	Number of Experiments.	Minimum Current.
		micro-ampere
Leaflet of *Mimosa*	25	0·7
" *Biophytum*	50	0·5
" *Neptunia*	5	7
" *A. carambola* ..	10	4
" *A. bilimbi*	5	6
Leaf of *Mimosa*	15	2

I next give a table showing results obtained with moderate current:—

TABLE XV.—EFFECT OF MODERATE CURRENT ON VARIOUS SPECIMENS OF SENSITIVE PLANTS

Type II.—Excitation formula **Km Ab**

Specimen.	Number of Experiments.	Minimum Current.
		micro-amperes
Leaflet of *Mimosa*	25	1·6
„ *Biophytum*	50	2·1
„ *Neptunia*	5	12
„ *A. carambola* ..	10	11
„ *A. bilimbi*	5	16
Leaf of *Mimosa*	15	4

It is here shown that we have examined not one species only but all the species that were available for experiment. More than a hundred experiments have moreover been carried out on different specimens. The result of all these investigations has been to demonstrate clearly the fact that in plants, under a feeble current, it is the kathode alone that excites at make, while under a moderate current excitation takes place by the make of kathode and the break of anode.

Thus, so far from there being any necessary divergence in the excitatory polar reactions of fibrillated and unfibrillated protoplasm, we find on the contrary that they are identical in the plant and in the animal. In other words, the responsive characteristics of Type I. and Type II. in plants correspond in every respect to those in animal tissues which are classed under Pfluger's Law as Stage I. and Stage II.

This is not by any means to be taken as a complete statement of the law of polar effects. We shall find that if the current be increased still further, certain new types of effects make their appearance which will be described in detail in a subsequent chapter. In the meantime I have to describe certain further characteristics of these polar effects and the relative sensitiveness of plant and animal to this form of stimulation. These will be dealt with in the following chapters.

SUMMARY

The polar effects observed in diverse sensitive plants are in every way similar to those in animal tissues. THE LAWS OF POLAR EXCITATION IN PLANTS ARE :—

I. With feeble current the kathode excites at make and not at break. The anode excites at neither make nor break.

II. With current of moderate intensity the kathode excites at make and not at break. The anode excites at break and not at make.

CHAPTER XVI

THE CONTRASTED EFFECTS OF ANODE AND KATHODE

Polar effects of currents on pulsation of *Desmodium gyrans*—Reduction of systolic contraction by anodic action—Diminution of diastolic expansion by kathodic action—Arrest at systole by make of kathode and diastolic expansion by break of kathode—Arrest at diastole by make of anode, and systolic contraction by break of anode—Effects of ascending and descending currents of feeble and strong intensity in nerve-and-muscle preparation—Parallel effects in petiole-and-pulvinus.

WE have seen that under currents of moderate intensity the polar effects of anode and kathode are more or less antithetic. The kathode excites at make, whereas the anode excites at break. If the kathode at make induces contraction, then it would appear probable that at the make of anode there may be produced an expansion. At break, again, supposing the same antithesis to be maintained, the contractile effect at the anode will have a contrasted effect of expansion at the kathode.

Biedermann has shown that in a beating heart the point of application of the anode, during make, remains expanded as a dark red blistered swelling, even when the rest of the heart, during contraction, is becoming pallid. This shows that during the continuation of the anode an expansive reaction is induced in the tissue. An effect similar to this is induced at the kathodic point at break.

I have been able to demonstrate the contrasted effects of anode and kathode, in a still more striking manner, in the case of rhythmic pulsation of the leaflet of *Desmodium gyrans*.

In a future chapter it will be shown how remarkable are the similarities of the rhythmic tissue of the plant and animal. In the case of the *Desmodium* leaflets the pulsations are seen to take place in a very regular manner, the period of one complete pulsation being about 3 minutes. Of the up and down movements of the leaflet, the down movement takes place more quickly, corresponding to the systolic contraction of the rhythmic cardiac tissue. The systolic movement in *Desmodium* takes place in the course of about 1 minute and 10 seconds, the slower diastolic expansion being accomplished in the course of about 1 minute and 50 seconds. In the record, the quicker systolic movement is represented by the up-curve. The extent of contraction is thus represented by the upper limit of the curve of response; the lower limit indicates the extent of diastolic expansion. If the leaflet is executing its greatest possible amplitude of pulsation, then an external agent will be unable to increase it any further; but the extent of contraction or expansion can be individually reduced under agencies which have an opposing tendency. Thus if the continuation of anode tends to induce expansion, then during a cycle of pulsating activity it would oppose the contraction at the systolic phase. The effect would be a diminution of contraction; in the record this would appear as progressive diminution of heights of responses.

If the application of kathode, on the other hand, induces contraction, this would oppose the diastolic expansion; the amplitude of pulsation will be progressively diminished, with continuously diminishing relaxations, and the base-line would be shifted upwards.

In order to demonstrate these contrasted polar reactions the experiment was carried out by making the pulvinule of the leaflet alternately anode or kathode. One electric connection is made with the pulvinule by means of a thin thread, special care being taken that this in no way interfered with the free pulsation of the leaflet; the second connection is made with an indifferent point lower down in the petiole.

I shall first describe the expansive effect of anode-make and the contractile effect of kathode-make; these particular effects are easy to demonstrate, the current applied being of moderate intensity. In fig. 116 is seen the anodic effect; the first two pulsations are normal, after which the anode was applied. This is seen to result in a continuous lessening of the height of successive contractions.

The application of kathode, on the other hand, is found to induce a precisely opposite effect (fig. 117); here the diastolic expansion is opposed, which results in a continuously diminishing relaxation. In fig. 116 a line joining

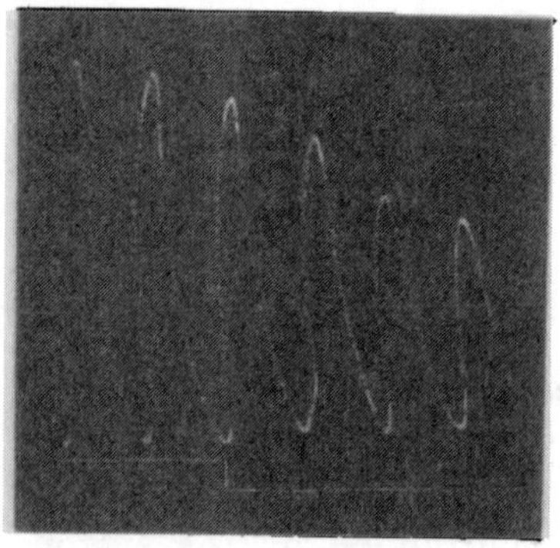

FIG. 116.—Effect of anode, opposing contraction in pulsation of *Desmodium gyrans*. Note the gradual diminution of systolic contraction.

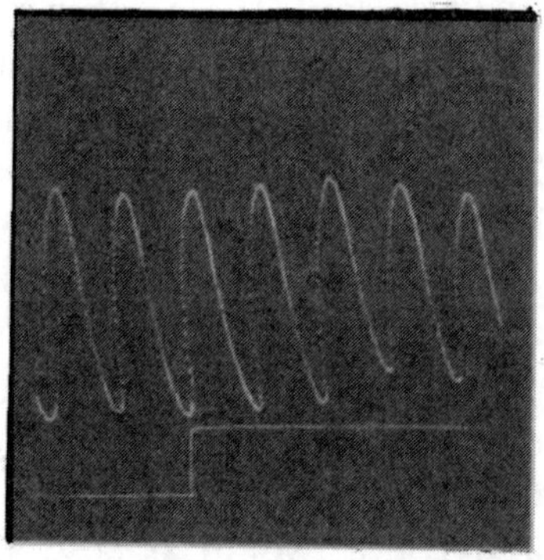

FIG. 117.—Effect of kathode, opposing expansion. Note the gradual diminution of diastolic expansion.

the apices of successive contractions is seen to descend, while in fig. 117, under the action of kathode, the line joining the extreme points of the diastolic excursion is seen to ascend. The same fact is seen again in fig. 118, where a single specimen is subjected first to anode and then to kathode. The contrasted effects of anode and kathode in this case are very obvious.

By employing a suitable intensity of current it is sometimes possible to exhibit the contrasted effects of the kathode and anode by the actual arrest of pulsation. By the make of kathode the arrest takes place towards systole, and by the make of anode towards diastole. In such cases it is possible to demonstrate further the remarkable effects of the break of kathode and break of anode. The difficulty in this experiment lies in the fact that the application of an intensity of current, greater than is exactly sufficient to induce the arrest, is apt to bring about fatigue, with the abolition of pulsation. I have several times succeeded in

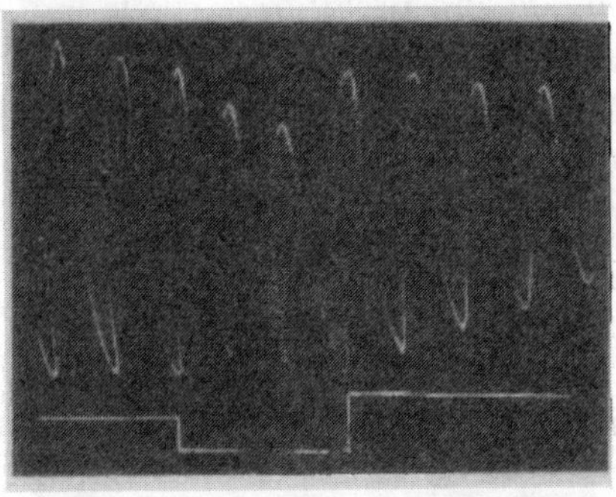

FIG. 118.—Alternate effects of anode and kathode in diminishing systolic contraction and diastolic expansion.

inducing an arrest which was not followed by a permanent abolition of excitability. In these cases it is possible to exhibit the very interesting effects of break of kathode and of anode. In fig. 119 is seen the arrest at systole induced

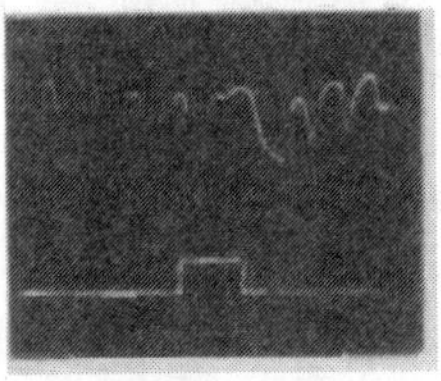

FIG. 119.—Arrest at systole by the make of kathode and diastolic expansion as immediate effect of break of kathode.

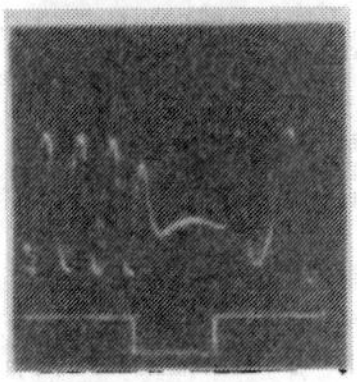

FIG. 120.—Arrest at diastole by the make of anode and systolic contraction as immediate effect of break of anode.

by the make of kathode. The current was then broken, and as an immediate result an expansive or diastolic phase of pulsation was obtained, followed by several other pulses of somewhat diminished amplitude. The effect of break of anode, on the other hand, is exactly opposite. In fig. 120

the pulsation was arrested towards diastole by the make of anode. The break of anode was immediately attended by an abrupt termination of the standstill. A pulse of systolic contraction followed, and was succeeded by the renewal of ordinary pulsation. It is thus seen that while the make of kathode and break of anode induce contraction, the make of anode and break of kathode induce the reverse effect of expansion.

The contrasted effects of anode and kathode, and of make and break, are also exhibited by induced variations of excitability. These are well seen in characteristic differences of effects induced by a constant current in a muscle-and-nerve preparation, these being modified by the strength of current. Thus employing a feeble current it is found that both ascending and descending currents in the nerve induce excitation of the terminal muscle at make but not at break. With strong currents, on the other hand, excitation only takes place at the make of the descending current and break of the ascending current. We shall presently see how these different effects are explicable on the assumption of the contrasted effects of anode and kathode, and of make and break.

The obscurity of the subject lies in the fact that there is no visible indication of an excitatory change in the nerve. The case would have been different had the conducting-tissue been provided with indicators which could signal the passage of excitation. Such a conducting-tissue is provided by the petiole of *Biophytum*, where the successive closures of the lateral leaflets indicate the transmission of excitation through the central conducting-strand. We shall see how the effects visually manifested in *Biophytum* elucidate the characteristic effects observed in the nerve-and-muscle preparation.

Fig. 121 shows arrangements of parallel experiments in nerve of frog and conducting petiole of *Biophytum*. The polarising electrodes are applied to the right, the terminal motile organ, muscle or leaflet, being to the left. The series to the left represent the effect of descending, those to the right the effect of ascending, current. In the case of the descending current the terminal indicator is to the left of K, and the conducting tissue between K and A. In the case of the ascending current the terminal indicator is to the left of A. Transmitted excitation can be detected by the

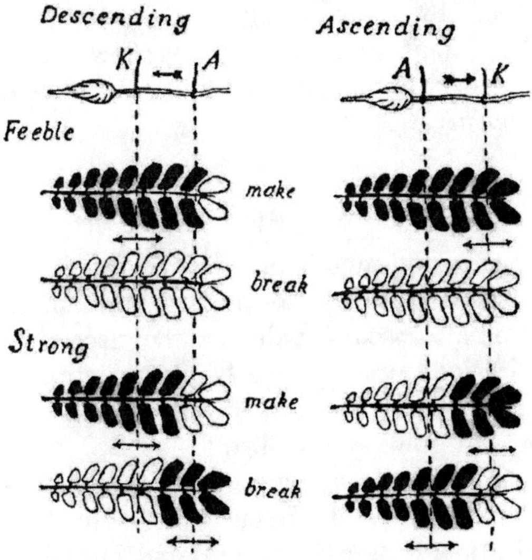

FIG. 121.—Effects of Descending and Ascending currents at make and break on nerve-muscle and petiole-pulvinus. Excited leaflets shaded dark. Vertical series to left represent effects of descending currents; series to right effects of ascending currents. With feeble current, terminal leaflets excited by both descending and ascending currents at make and not at break. With strong currents, excitation of terminal leaflets takes place only at make of descending and break of ascending currents. Two upper pairs of leaves exhibit effects of feeble current at make and break. The two lower pairs of leaves show effects of strong current.

contraction of muscle or leaflet, to the left of K in the case of descending, and to the left of A in the case of ascending current. But the characteristic effect at the electrodal points, K and A themselves, or the overflow of the effect between the two points, cannot be detected in the case of frog's nerve. It can, however, be detected in the case of conducting-tissue of *Biophytum* on account of the presence of the lateral motile-leaflets.

I will now describe experiments carried out on *Biophytum* with feeble and strong currents, and show how they throw light on parallel experiments with nerve-and-muscle preparation of animal. In *Biophytum* the excited leaflets are indicated by dark shading, which will also clearly exhibit the extent of excitatory overflow. The point of initiation of excitation, and direction of transmission, are indicated by arrows.

Effect of Feeble Current

In the nerve-and-muscle preparation it is known that (1) excitation takes place at the make of descending current; (2) excitation also occurs at the make of ascending current; (3) no excitation occurs at the break of descending, (4) or ascending, currents. Accounts of parallel experiments with *Biophytum* will now be described :—

(1) *Make of descending current.*—The kathode is proximal to the extra-polar or terminal leaflets to the left. At make, excitation is seen to be initiated only at K, the two waves proceeding in opposite directions. Five pairs of leaflets thus fall in the extra-polar and three pairs in the intra-polar regions. This excitatory overflow in the intra-polar region cannot be observed in the animal nerve on account of absence of a visible indicator.

(2) *Make of ascending current.*—In fixing our attention to the terminal or extra-polar leaflets to the left of A, we find occurrence of excitation corresponding to the excitation of muscle at the make of the ascending current. That the excitation was really initiated at the distal kathode, and traversed without hindrance through the feeble anode, will clearly be seen from the serial closure of the leaflets initiated at K. Eight pairs of leaflets underwent excitatory fall to the left, and one pair to the right of K.

(3 and 4) *Break of ascending and descending currents.*—

There is no excitation, since feeble anode does not excite at break.

Effect of Strong Current

We now take up the question of the effect of strong currents :—

(1) *Make of descending current.*—In the nerve-and-muscle preparation there is excitation of muscle at the make. Here the kathode is proximal, and we obtain the normal excitatory effect of kathode-make.

In the corresponding experiment with *Biophytum* we find excitation transmitted to the terminal or extra-polar region to the left, five pairs of leaflets undergoing closure. Only two pairs of leaflets closed in the intra-polar region, further progress of excitation being arrested by the depressing action of the anode.

(2) *Make of ascending current.*—Unlike the action of feeble current, there is no excitatory effect in the nerve-and-muscle preparation at the make of strong ascending current. This is explained on the supposition that the excitation at the distal kathode cannot traverse the region of the strong anode with its depressed excitability. The proof of this assumption is strikingly afforded by the corresponding experiment with *Biophytum.* We observe the initiation of excitation at the kathode, but the progress of excitation is arrested near the region of anode. Hence, in spite of the occurrence of excitation, there could be no transmitted effect in the terminal extra-polar region.

(3) *Break of descending current.*—In a nerve-and-muscle preparation there is no excitation at break of a descending current. Here, though the distal anode excites at break, the intervening kathodic region is assumed to undergo depression at break. Hence a block occurs to the transmission of excitation.

This is made quite clear in the corresponding experiment

with *Biophytum*. We here observe excitation initiated at the anode at break, two pairs of leaflets undergoing closure, and the further progress of the excitatory wave is arrested before reaching the depressed region of kathode-break.

(4) *Break of ascending current.*—Excitation occurs in a nerve-and-muscle preparation at the break of strong ascending current. Here the anode is proximal, and excitation induced at break reaches the terminal organ without hindrance.

In the corresponding experiment with *Biophytum* we observe five pairs of leaflets undergoing closure in the terminal extra-polar region, the excitation being initiated at the anode at break. Excitation also traversed the intra-polar region, two pairs of leaflets undergoing closure. Further progress in this direction was, however, arrested by the depressing action of kathode-break.

SUMMARY

The contrasted effects of anode and kathode are exhibited by appropriate modification in the pulsating activity of *Desmodium gyrans*.

The anodic effect of expansion is seen in the reduction of normal limit of systolic contraction.

The kathodic effect of contraction is observed in the reduction of normal limit of diastolic expansion.

The immediate effect of break is the reverse of that at make or continuation of current.

The diastolic arrest by anode is followed at its break by systolic contraction.

The systolic arrest by kathode is succeeded at its break by diastolic expansion.

In nerve-and-muscle preparation the effects of ascending and descending currents are found modified by the intensity of the current. Effects in every way parallel are observed in experimenting with petiole-pulvinus of *Biophytum.*

These characteristic modifications are easily traceable in *Biophytum* to the contrasted effects of anode and kathode, and of make and break. Excitability is enhanced by the make of kathode and break of anode. It is depressed by the make of anode and break of kathode.

CHAPTER XVII

EFFECT OF TEMPERATURE ON POLAR EXCITATION, AND MULTIPLE EXCITATION UNDER CONSTANT CURRENT

Excitability of conducting tissue to induction-shock diminished by cooling—Nerve-excitation by constant current enhanced by cooling—Excitation of conducting-tissue of *Mimosa* by constant current enhanced by cooling and depressed by warming—Ineffective stimulus becoming effective under cooling and *vice versa*—Multiple response induced in *Biophytum* by the passage of constant current—Comparison of sensitiveness of plant and animal—Minimum current for excitation of human tongue—Relatively higher sensitiveness of *Biophytum*.

In continuing the investigation on polar reaction, certain modifications of effect were noticed as the season advanced from spring to summer. Some of these will be described in a succeeding chapter. But one modifying effect—namely, that due to temperature—appeared at first very puzzling.

The temperature of Calcutta in summer is high. Moreover experiments with plants had to be carried out in a glass-house. Thus the temperature to which the plants were subjected in summer was often as high as 35° C. In these circumstances it was found that the indirect stimulation of the pulvinus of *Mimosa* by the action of constant current often became ineffective. It was quite easy in spring to excite the leaf of *Mimosa* by the transmitted excitation due to the make of kathode when the kathodic point was at a distance of several centimetres from the pulvinus. But in summer there was hardly any transmitted excitation even when the exciting kathode was at a comparatively short distance from the pulvinus.

This ineffectiveness might be due to the impairment of conductivity or excitability. It could not be due to the loss of conductivity, for we have seen in a previous chapter that the conductivity is enhanced by a rise of temperature. The pulvinus, again, was found extremely sensitive. Yet, in spite of the high conductivity and motile excitability, the transmitted effect of excitation was often found ineffective at a high temperature.

Thus the only remaining factor to which the change might be attributed was the excitability of the conducting tissue itself. Taking the parallel case of the animal nerve, it is known that the excitability of a nerve is diminished by local cooling, the stimulus being that due to break induction-shock. But Gotch and Macdonald have made the very interesting observation that the effect is reversed in the case of stimulation by constant current ; here the nerve excitability is enhanced by lowering of temperature. By employing a descending current—with the kathode proximal to the contractile muscle—they found that the make-excitation which was ineffective when the nerve was locally warmed to 30° C. became effective when locally cooled to 5° C.

It occurred to me that the failure of indirect stimulation by the closure of constant current in *Mimosa* might be due to the depression of excitability of the conducting petiole, in consequence of the high temperature. Should this prove to be the case, then this specific reaction would afford a very striking demonstration of the characteristic similarities in the conducting-tissues of the animal and plant.

The effect of cold in modifying the excitability of the conducting animal nerve is, as said before, dependent on the mode of stimulation. We shall now see whether this holds good in the case of the plants also. First, in order to determine the effect of cold on the exciting efficiency of the break induction-shock, the two electrodes from the secondary coil were placed on the petiole, one 20 mm. from the pulvinus and the other further away. It has been shown that with a single induction-shock it is the kathode which causes excitation. The electrodes of the secondary coil are so connected with the petiole as to render the proximal contact the kathode, and therefore the point of excitation. Records are then taken under the same stimulus alternately, (1) with the excited point as warm as the general temperature of the room, which was 30° C., and (2) with the temperature lowered to about 5° C. by the application of cooled water. It is essential that the cooling should be effected gradually, for sudden variation of temperature of itself causes excitation.

The series of records in fig. 122 shows the result. It is seen that while the excitation is effective at H, H, H, when

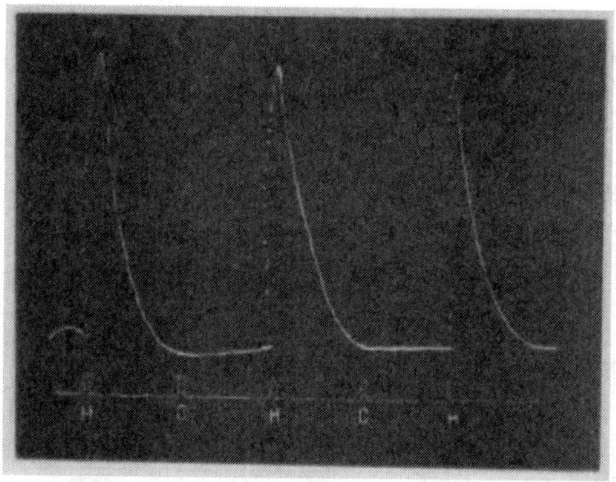

FIG. 122.—Effect of cold on excitability to induction-shock : H, C, alternate effective and ineffective excitation at moderately warm and low temperatures respectively. Testing stimulus, a single break induction-shock was maintained constant.

the stimulated point is warm, it becomes ineffective at C, C, when the excited point is cooled.

This proves that in the conducting-tissue of the plant lowering of temperature depresses the excitatory efficiency of a break induction-shock.

For studying the effect of cold on the exciting efficiency of the constant current, I next made suitable electrical connection with two points on the petiole, the proximal kathode, K, being at a distance of 10 mm. from the pulvinus. Excitation was produced by the make of the descending current. An E.M.F. was applied which caused minimal response; this was found to be 2 volts, the temperature of the room being 30° C.

The point K was now alternately raised and lowered in temperature. This was effected by means of a stream of hot or cold water applied at the point. It is essential that the warming or cooling should be effected gradually, for, as stated above, sudden variation of temperature of itself causes excitation.

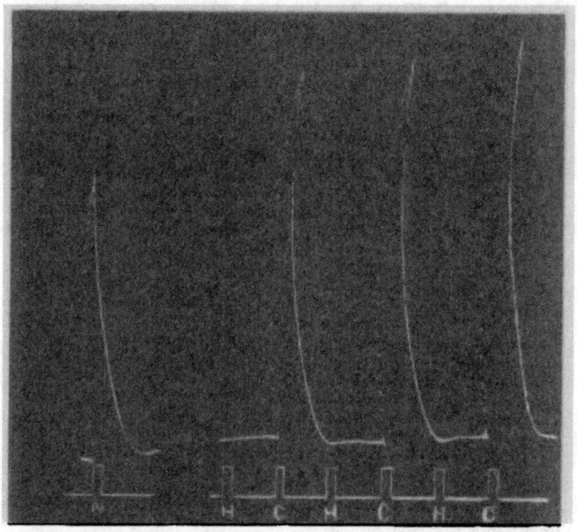

FIG. 123.—Record showing abolition of polar excitation at high temperature: N, normal response; H, C, alternate ineffective and effective excitations at high and low temperatures respectively. Testing stimulus of kathode-make was kept constant.

A series of automatic records are shown in fig. 123, made by the plant. The first record of the series, N, gives the response to the descending make-current; the amplitude of this is only moderate. The petiole was locally warmed to about 37° C., and an identical stimulus of the make of descending current was applied at a moment marked H. It will be noticed that the stimulus which was formerly effective has now become ineffective.

The petiole was next locally cooled to about 10° C. and

stimulus of descending make current applied as before at C. It will be seen that not only is the stimulus now effective, but the amplitude of response is much greater than that of normal response at 30° C. In order to test these results further, the experiment was repeated under alternate heating and cooling. It will be noticed that under rising temperature there is an invariable failure of excitation, while under lowered temperature excitation occurred always and became maximal. Therefore one of the features which characterise the animal nerve is found also in the conducting-tissue of the plant.

Multiple Excitation by Constant Current

A very striking result of the passage of a constant current is the production of multiple responses in certain rhythmic tissues. Thus the non-ganglionated preparation of the apex of the frog's heart is under ordinary conditions quiescent, but on the passage of a constant current it breaks forth into a rhythmic series of excitations.

In a subsequent chapter it will be shown that certain vegetable tissues exhibit multiple responses. The pulvinus of the leaflet of *Biophytum* is multiple-responding and behaves in many respects like the cardiac tissue. Now on maintaining a constant current of certain intensity through the petiole, the kathode being placed on a pulvinus, the particular leaflet will be found to execute a series of pulsatory movements.

Of still greater interest is the indirect effect of such stimulation. A particular leaflet of *Biophytum* is attached to the recording lever. A constant current is sent for a short time through the petiole, the kathode being at a distance of 10 mm. from the leaflet. The leaflet responded to the excitation caused by kathode-make; the response here was due to the excitation transmitted through the distance of 10 mm. The first part of fig. 124 gives four sets of responses to individual stimuli, applied at intervals of 3 minutes. After this the petiole was subjected to the action of a continuous current, represented below the record as a continuous up-line. It will be seen that under the continuous current, rhythmic excitations were initiated at the kathode, which, reaching the responding leaflet, caused multiple responses.

There were five such rhythmic responses in the course of 10 minutes, the period of each being 2 minutes. The

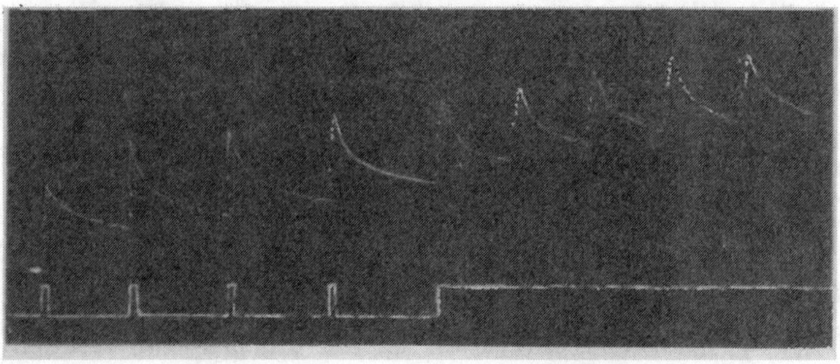

FIG. 124.—Multiple excitation in *Biophytum* under constant current. First four records are responses to individual stimuli of electrical current of short duration, applied at intervals of 3 minutes. The last five are multiple responses due to continuous action of current.

production of such multiple excitations in a plant under constant current cannot be explained by the theory of hydro-mechanical blow. On the other hand, their similarity to corresponding phenomena in animal tissues is sufficiently obvious.

RELATIVE SENSITIVENESS OF PLANT AND ANIMAL TO ELECTRICAL CURRENT

There is a general assumption that the sensitiveness of plant tissue is very much lower than the animal tissue. The question then arises: How sensitive is the plant as compared with the animal in the matter of excitation by a constant current?

There can probably be no specimen more susceptible

of such excitation than the nerve-and-muscle preparation of a frog. This preparation is an exceedingly sensitive detector of induction-current. A shock too feeble to be felt by the intact human subject when passed across the body through two fingers, will, when applied to the exposed nerve of the frog, provoke vigorous movement in the attached muscle. The intact plant, like the human subject, is not so sensitive to an induction-shock as the bare nerve.

But the case is different when we come to the question of excitation by a constant current. In order to compare the relative excitatory effects here, in plant and animal, I carried out two different investigations, comparing the sensitiveness of the plant on the one hand with that of frog's nerve-and-muscle preparation, and on the other, with that of intact human subject.

In the first of these I placed an intact petiole of *Biophytum* in series with a nerve-and-muscle preparation of frog. A gradually increasing E.M.F. was applied by means of a potentiometer-slide arrangement, till one of the two specimens showed excitation at make. The intensity of the exciting current was now measured by means of micro-ammeter. The E.M.F. was then further increased till the second of the two specimens gave excitation at make, and the current again measured. From the value of these two effective currents the relative sensitiveness of the two specimens can be gauged. Instead of measuring the current, we may take, if we wish, the readings of the slide-wire for the two effective values.

These experiments were carried out in August, during the rainy season. As regards the specimen of *Biophytum*, it should be mentioned that the vitiated atmosphere of a town is particularly inimical to the sensitiveness of this plant. I have succeeded in raising *Mimosa, Desmodium*, and other sensitive plants in Calcutta, without much loss to their sensitiveness; but I have invariably failed to do this in the case of *Biophytum*. It always becomes stunted and discoloured, and ceases to exhibit its normal motility. The only alternative has been to grow the plant in the suburbs and bring it to the laboratory. Here, with one or two days' rest after the disturbing effect of transport, it regains a fair degree of sensitiveness, but the effect of the air of the town is to bring about a daily deterioration, and in the course of a week or ten days, except in rare cases, it becomes practically insensitive. It will thus be seen that the experimental plant employed for our comparison may be taken as having had its sensitiveness reduced certainly to half.

In proceeding with the experiment, according to the method already described, it was found that the nerve-and-muscle preparation responded at the make of a current which was as feeble as ·25 micro-ampere. The *Biophytum* at this time had not yet responded, but when the current intensity was increased to ·5 micro-ampere excitation was seen to occur. From this it would appear that *Biophytum* had in this case fully half the susceptibility of the nerve-and-muscle preparation of frog. Bearing in mind the peculiarly unfavourable character of the circumstances to which the plant was in this case subjected, it does not seem too much to infer that the sensitiveness of the two specimens was not naturally of a different order.

It might perhaps be still more interesting to institute a comparison between the intact plant and the intact human subject. The tongue has always been considered a very sensitive detector of electrical current. According to Laserstein, the acid effect of anode is appreciated by it when the current is only $\frac{1}{156}$ of a milliampere. This is equal to 6·4 micro-amperes.

But I have found amongst my pupils some who could perceive by the tongue a current as feeble as 1·5 micro-ampere, whereas the *Biophytum* in the same circuit responded to the much smaller current of ·5 micro-ampere only. This demonstrates that the plant was in this case three times as sensitive as the human tongue. A more excitable specimen of *Biophytum* would doubtless prove to be about ten times as sensitive.

The improbability of the theory of hydro-mechanical disturbance becomes evident when we realise that an excitatory impulse is initiated and transmitted in the plant under a stimulus that cannot even be perceived by the extremely sensitive human tongue.

SUMMARY

Nerve-excitation by constant current is enhanced by cooling and depressed by warming. This is one of the characteristics of polar excitation by constant current.

Similarly the excitability of conducting petiole of *Mimosa* to polar excitation is enhanced by cold and depressed by warmth. Minimal excitation becomes maximal under cold, and ineffective under warmth.

The passage of a constant current through the petiole of *Biophytum* gives rise to series of multiple excitations.

The sensitiveness of *Biophytum* to an electrical current is remarkably high. Compared with the very sensitive human tongue, the sensitiveness of *Biophytum* is about ten times as great.

CHAPTER XVIII

POLAR EFFECTS UNDER STRONG CURRENTS

Abnormal polar reactions in *Protozoa*—Transformation of polar reaction in leaf of *Mimosa* from Type II. to Type III. under strong current—Further transformation to Type IV. under stronger current—Exhibition of Type III. and Type IV. by leaflets of *Mimosa, Biophytum,* and *Averrhoa carambola*—Law of polar action of strong currents.

It has been shown in a preceding chapter that the characteristic effect of a feeble current in various plants was excitation at make of kathode, but that with a moderate current there was excitation not only at the make of kathode but also at the break of anode. These two types of effect corresponded to those recognised in the case of animal tissues under Pfluger's Law as Stage I. and Stage II. It will be remembered that in the case of *Protozoa* the polar reactions have been found to be very different from these. Thus Kühne found that in *Actinosphaerium* excitatory contraction took place at the make of both kathode and anode, and that excitation again took place at the break of kathode. The excitatory formula here, then, may be expressed as **Km Kb Am.**

It will be noticed that we have here the abnormal reactions of anode-make and kathode-break. Verworn has corroborated and extended these observations. He further finds that *Amphistigma* is excited by both anode and kathode at make. There was apparently no effect at break. The excitation formula in this case is therefore **Km Am.**

From these abnormal effects it has been suggested either that the polar effects in fibrillated and unfibrillated protoplasm are more or less opposed, or that there is no law of polar action which is of universal application.

But we have seen that the unfibrillated protoplasm of the plant exhibits polar effects which are identical with those of animal tissues. It is not impossible that normal polar effects may be subject to modification from the influence of diverse factors, such for example as strength of current, physiological condition of specimen, its age, and the season of the year. When an exhaustive survey has been made, it may be found that the responses of the *Protozoa* were not, after all, anomalous, but have a place of their own in some transitional type of reaction given by specimens whose characteristics are definitely recognised as normal. With the object of completing such a survey, we may continue our detailed study of the polar reactions of sensitive plants under widely different conditions. And first we shall study the effects of a still further progressive increase of current, in order to see whether any new phenomenon comes into the field of observation, taking the reactions of the pulvinus of *Mimosa* first in the series.

PRIMARY LEAF OF *MIMOSA*

I have employed both the bi-polar and mono-polar methods. The mode of procedure was to increase the current step by step from minimum to maximum, and note the changing types of reaction at different critical points. With vigorous specimens this can be done without any danger of the onset of fatigue. After the value of certain critical points has been determined, experiments were repeated with fresh specimens, at and about these particular critical points. I shall first describe a typical experiment with a very sensitive and vigorous specimen, and for the sake of simplicity I give in detail the effects observed at one pulvinus only, leaving it to be understood that similar effects were also induced at the other.

The specimen, as said before, was very sensitive, and

when the exciting current applied reached the value of 3·5 micro-amperes the characteristic effect of Type I.—namely, excitation at make of kathode **Km**—made its appearance. When the current was still further raised to 5·6 micro-amperes, the excitatory effect was of Type II.—namely, **Km Ab,** that is to say, excitation at the make of kathode and break of anode. In these two cases we have Types I. and II., with which we are already familiar.

POLAR REACTION TYPE III., **Km Am Ab.**

When the progressively increasing current had reached a value of 6·3 micro-amperes, however, a new and unexpected

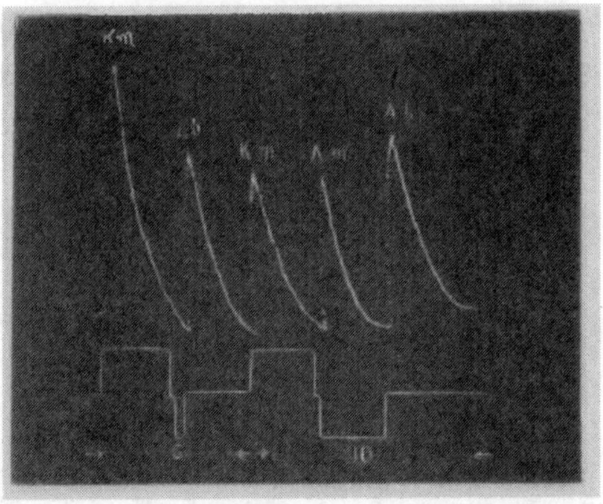

FIG. 125.—Effect of increasing intensity of current in transforming response **Km Ab** of Type II. to response **Km Am Ab** of Type III.

type of reaction made its appearance. With the bi-polar connections it was found that while one leaf, the kathode, underwent an excitatory fall, the other, or anode, also showed a simultaneous fall at make. After the re-erection of the leaves the current was broken, and now the anodic leaf showed excitation once more, whereas the kathodic exhibited no excitation whatever. In this new type of effect, then, we have excitation both at anode and kathode at make and at the anode at break only. This therefore constitutes Type III., with the excitatory formula **Km Am Ab.**

This type of excitation will be seen very clearly in the automatic record which is given in fig. 125, taken with a different specimen, where a single pulvinus was carried through a complete cycle of kathode-make, kathode-break, anode-make, and anode-break. It will be seen from the record that while under 6 volts excitation took place at kathode-make and anode-break, under a higher E.M.F. of 10 volts excitation not only took place at kathode-make and anode-break but also at anode-make. Under a stronger current Type II. has thus been transformed to Type III.

Polar Reaction Type IV., **Km Kb Am Ab**

Returning to the first specimen, the current was further increased till an intensity was reached of 12·7 micro-amperes,

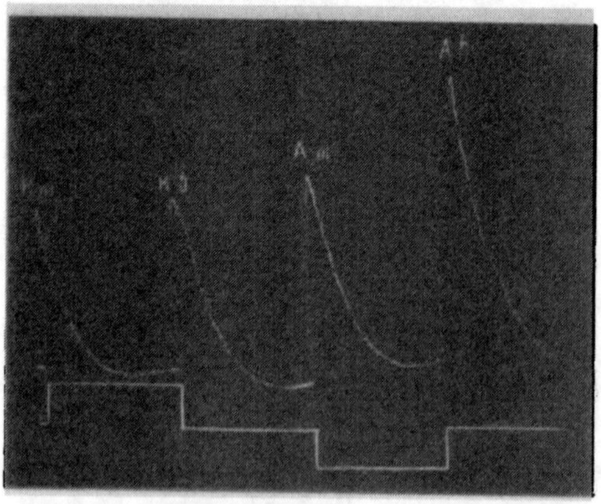

Fig. 126.—Polar excitation Type IV., **Km Kb Am Ab.**

and here was obtained another new type of reaction—namely, excitation at both kathode and anode at both make and break. The excitatory formula for this fourth type is thus **Km Kb Am Ab.**

In fig. 126 is exhibited an automatic record of this type

obtained with a given pulvinus when it was carried through the usual cycle of kathode-make, kathode-break, anode-make, and anode-break. It will be noticed that excitation took place here at each of these phasic changes.

In the case of the particular specimen whose consecutive changes have been traced, from the first type to the fourth, it will be noticed that it was subjected to a long-continued series of experiments. After this, however, I took up fresh specimens, with a view to the immediate observation in their case of Type III. and Type IV. There could here be no possible suspicion of changes induced by previous currents too protracted. Thus in a certain fresh specimen the third-stage effect was obtained with a current of 9·1 micro-amperes. When this was further increased to 12·6 micro-amperes the excitatory reaction was transformed into that of Type IV. It will be remembered that with less sensitive specimens a higher intensity of current is necessary for the induction of any particular type of effect. Thus with a certain less sensitive *Mimosa* a current of 14 micro-amperes was required to induce responses of Type III. The current had in this case to be increased to 32 micro-amperes before the effects were transformed to those of Type IV.

The particular excitations that appear extraordinary in these higher types are those at anode-make and kathode-break. The anode-make effect, as has been shown, can be induced in the initial response of a fresh specimen. But to obtain that at kathode-break—since this presupposes a previous kathode-make which, if sudden, has caused excitation—a certain time must be allowed to elapse after make, in order that the leaf under the continuous action of the current may re-erect itself, with restoration of its sensitiveness. Lest the long-continued action of the current should here have induced some unknown change to which the excitation at kathode-break might be attributable, I have carried out a number of test-experiments in which the kathodic increase of current was made gradually, though in a comparatively short time, without allowing any excitation to occur at make. The kathode was then broken suddenly, with the result that excitation was immediately exhibited. Hence the kathode-break effect under strong current must be taken as a normal phenomenon.

I have carried out more than fifty different sets of experiments on the characteristic effects of relatively strong currents, which have fully confirmed the results described. It was uniformly found that as the current was increased step by step, the transformation took place, as here laid down, to a higher type, and never consisted of reversion to a lower.

Taking these facts, then, as well established, the question arises whether the new, and apparently anomalous, effects cannot be explained in a simple way as a special case of these reactions with which we are already familiar? Thus the first explanation that occurred to me was that the fall of the leaf at anode-make might be due to the arrival of excitation at the anode from the distant kathode. Against this supposition, however, was the fact that a strong anode is known to be so depressing as to act as a block to the arrival of any excitation from outside. Overlooking this difficulty, the arrival of excitation from a very distant kathode on a different branch of the plant might be expected to take a certain interval of time. But in the cases given, the excitation at make of anode was, to all seeming, instantaneous. Finally, I killed a point on the distant branch by severe scorching, and made it kathode. By this means the possible excitatory effect at kathode was completely eliminated. In spite of this abolition of kathodic action, however, the excitation at anode-make continued to take place as before. Another modification of the experiment lay in previously killing the end of the petiole whose pulvinus was being experimented on. This was done, as before, by scorching. The intense excitation caused by this injury induced an excitatory action at the pulvinus, and it was only after the expiration of about an hour that the organ regained its excitability, and not even then to the fullest extent. The kathode was placed on the injured point, and the anode at or near the pulvinus; the exciting effect was again obtained at the make of anode. The only difference in this case lay in the fact that, owing to the partial loss of sensibility at the pulvinus, the intensity of current for inducing one of the higher types of response had to be increased.

From these experiments the new types of effects, Type III. and Type IV., appear to be normal. A further consideration of this question will be postponed till I have given in detail the reactions of strong currents with other species of sensitive plants.

Leaflets of *Mimosa*

I have already mentioned the misgiving I at first entertained that the somewhat unexpected appearance of excitation at anode-make should in reality be due to transmission from the distant kathodic-point. I have also described the several means by which these doubts were finally set at rest, one of those being previous injury of the kathodal-point by scorching, in order to eliminate it as a possible source of excitation. I was still desirous, however, of finding some independent means of demonstrating conclusively the independent character of the excitation under a strong current at anode-make.

For this purpose I found the leaflets of *Mimosa* eminently suitable. The two bi-polar connections are made at about the middle points of two different sub-petioles, care being taken that the contact of each electrode is made on the midrib at the insertion-points of two opposite pulvinules. With this arrangement the character of excitation at anode-make, whether independent or transmitted effect from the distant kathode, can quite easily be discriminated, for it is impossible that any excitation should be transmitted along one sub-petiole to another without causing the fall of the intervening leaflets. In the case of independent excitations the fact will be displayed by the immediate fall of leaflets at the particular points, in this case the two electrodes. As regards the transmitted effects of excitation, these can easily be followed by watching the serial fall of leaflets.

I shall now describe an experiment in which, the connections being as already described, the current was increased step by step to a maximum. Beginning with a current of ·75 micro-ampere, it was found that excitation took place only at the kathode-make. When the current had been increased to 2·5 micro-amperes, excitation was initiated at the kathode at make and at the anode at break. In these two cases we have the familiar effects of Type I. and Type II. With the same identical specimen, however, when the current was raised to 5 micro-amperes it was found that excitation was simultaneously initiated at the bi-polar contacts, both kathode and anode, at make. The fact that the anodic fall had nothing to do with any transmission of excitation from the kathode was evident by the slow march of the excitatory waves, sent out from the two points independently, towards each other. It may be mentioned here that this phenomenon can be made still more prominent by selecting for experiment leaflets which are a little older and not excessively sensitive. In such a case the march of the two waves can be made as slow as may be desired.

We have seen that with strong current there is excitation at both kathode and anode; on the break of the current, excitation was now found to take place only at the anode. Thus we have the excitatory effect of Type III.—namely **Km Am Ab**—established independently by an experiment on the leaflets of *Mimosa*.

Employing the minimal current essential to the manifestation of Type III. we find the excitation at anode-make somewhat localised, whereas that at kathode-make is transmitted to a considerable distance. But on increasing the current the power of excitatory transmission of anode-make is greatly enhanced.

Returning once more to the experiment, I found that

when the current had been increased to 12 micro-amperes, excitation took place at both kathode-make and kathode-break and also at anode-make and anode-break. Here we have the characteristic effects of Type IV.—namely, **Km Kb Am Ab.** The following table gives a synopsis of these results :—

TABLE I.—EFFECTS OF CURRENTS OF INCREASING INTENSITY ON LEAFLETS OF *MIMOSA*

Intensity of current.	Excitation induced.	Characteristic type.
micro-amperes		
·75	**Km**	Type I.
2·5	**Km Ab**	Type II.
5	**Km Am Ab**	Type III.
12	**Km Kb Am Ab**	Type IV.

These conclusions were uniformly supported by the results of no fewer than thirty different sets of experiments on the leaflets of *Mimosa*. The only variations between different experiments lay in the fact that with less sensitive specimens a relatively higher current was required. Thus in a leaf which was somewhat older, and therefore less sensitive, the intensity of the current necessary for Type I. was 1 micro-ampere; for Type II. 3 micro-amperes; for Type III. 7·4 micro-amperes; and for Type IV., 15 micro-amperes.

LEAFLETS OF *BIOPHYTUM*

With leaflets of *Biophytum* I obtained results which were practically the same as those given by leaflets of *Mimosa*. Thus with a given experimental specimen, the characteristic effect, **Km** of Type I. was exhibited under a current-intensity of ·5 micro-ampere. When the current was increased to 2 micro-amperes the response was transformed to **Km Ab,** that of Type II. At 3·5 micro-amperes the characteristic effects of Type III.—**Km Am Ab**—made their appearance.

And finally, with a current of 10 micro-amperes the typical response of Type IV. was obtained—namely, **Km Kb Am Ab.** These results are shown in the following tabular statement :—

TABLE II.—EFFECTS OF CURRENTS OF INCREASING INTENSITY ON LEAFLETS OF *BIOPHYTUM*

Intensity of current.	Excitation induced.	Characteristic type.
micro-amperes		
·5	**Km**	Type I.
2	**Km Ab**	Type II.
3·5	**Km Am Ab**	Type III.
10	**Km Kb Am Ab**	Type IV.

These results were confirmed by fifty sets of experiments on different specimens of *Biophytum.*

LEAFLETS OF *AVERRHOA CARAMBOLA*

These leaflets are, as already explained, relatively insensitive and therefore require a higher E.M.F. with higher current. With these specimens also I obtained, as before, the four types of effects in their usual sequence. Thus in a given experiment a current of 4 micro-amperes gave **Km,**

TABLE III.—EFFECTS OF CURRENTS OF INCREASING INTENSITY ON LEAFLETS OF *AVERRHOA CARAMBOLA*

Intensity of current.	Excitation induced.	Characteristic type.
micro-amperes		
4	**Km**	Type I.
11	**Km Ab**	Type II.
20	**Km Am Ab**	Type III.
30	**Km Kb Am Ab**	Type IV.

the effect indicative of Type I. ; when the current was now raised to 11 micro-amperes excitation at **Km Ab,** constituting

Type II., was obtained. With a current of 20 micro-amperes the result was **Km Am Ab,** or Type III. And finally, with 30 micro-amperes the characteristic reaction of Type IV.—namely, **Km Kb Am Ab**—was observed.

These results were confirmed with ten different specimens.

I have thus described about a hundred experiments with different species of sensitive plants which conclusively demonstrate the existence of Types III. and IV. of polar reaction.

There remains an alternative hypothesis in regarding the effects seen in Types III. and IV., as in some way due to the production of secondary poles. The exact physical conditions under which the formation of the secondary pole is possible so as to cause complications in the excitatory phenomena are clearly shown in the experiments of Engelmann and Biedermann on the ureter of rabbit. In an insulated specimen, under a moderate current, it was found that excitation took place only at the kathode at make and at anode at break. But when the specimen was laid on a good conducting support, such as salt clay, then the polar reactions were found to be exactly reversed—that is to say, the anode excited at make and the kathode at break. This opposition of effects under differing circumstances is explained by the fact that in the second of these cases we have a conducting-sheet which gives rise to diffusion of the current and a rich development of secondary poles. Thus, opposite to the kathode, there are produced numerous secondary anodes, and conversely, opposite to the anode, there are numerous secondary kathodes. It is these secondary kathodes which are effective in causing excitation in the neighbourhood of the anode at make. In the neighbourhood of the primary kathode, on the other hand, excitation is prevented by the depressing influence of the secondary anodic points. The presence of secondary poles thus induces an apparent reversal of the normal effects, which is simultaneous at the two electrodes. The simultaneity of the reversal is therefore a presumptive evidence for the occurrence of secondary poles.

It may be asked next whether the effects seen in *Mimosa* as Types III. and IV., opposed as these are to Pfluger's Laws, could be explained away by any of the physical conditions of the experiment. In answer to this it may be pointed out that in the experiments described, which were carried out according to the bi-polar method, the electrical connections were made directly on the two contractile pulvini themselves, the thread forming a complete loop round each organ, which was thus equally and throughout its circumference anode or kathode, as the case might be. The organ in this case, moreover, was isolàted and free, like the insulated ureter. Hence there was an absence of all those conditions which might favour the formation of secondary poles. Had there been any such possibility we should have expected to see the reversal of normal effects, more or less, from the beginning, and the reversal should have occurred at the two bi-polar contacts simultaneously.

Instead of this, we found on the contrary that in the first two stages, with feeble and moderate currents, the reactions of the organ were absolutely normal. It was only after this, with the same specimen and with identical connections, that by merely increasing the current we obtained the effect characteristic of the third type—namely simultaneous excitation of kathode and anode at make, and excitation of anode at break. Had there been any induction of secondary pole, the result would have been excitation at anode at make and at kathode at break.

From these considerations, in addition to others already given, it appears that the normal effect of strong currents on *Mimosa* and other sensitive plants is to cause excitation at both kathode and anode at make, and anode only at break. These results, and those of Type IV. which immediately succeed, are definite and different from either of the two types that had previously occurred. These characteristic excitations, moreover, cannot very well be explained on the assumption of hypothetical secondary poles. The view that these effects are physiological will become strengthened when in the course of the next chapter we shall observe the modification of polar effects under physiological changes.

SUMMARY

As moderate increase of intensity of current transforms the polar reaction from Type I. to Type II., so also further increase in the intensity of current gives rise to reactions indicative of other types.

Pfluger's Law is not a complete statement of the polar action of currents. Under strong intensities of current, two additional types of reaction—Type III. and Type IV.—make their appearance. These are included in the following supplementary LAW of POLAR EFFECT of STRONG CURRENTS :—

Under the action of strong current, excitation takes place at the make of kathode and make and break of anode.

Under still stronger currents, excitation takes place at the make and break of both kathode and anode.

These four characteristic types of reactions are serially observed in all sensitive plants under increasing intensities of current.

CHAPTER XIX

VARIATION OF POLAR REACTION UNDER TISSUE MODIFICATION

Modification of polar reaction under tissue-changes—Effect of age—After-effect of moderate stimulation—Modified polar effect: excitation, at kathode-make and anode-make; excitation at kathode-make, kathode-break, and anode-make—General review of polar reactions.

In the course of my investigations into the effects of currents on the pulvini of different leaves of *Mimosa*, I found that during the beginning of the summer season the normal effects which have been described occurred uniformly, in their proper sequence. But later in the year, at the end of the rainy season, when the plants had begun to seed, I was puzzled by the appearance of certain new and unexpected types of response with which I had not hitherto been familiar. Thus in the case of certain leaves, with a moderate current the excitatory reaction took place only at make of kathode and make of anode. It will be remembered that in *Amphistigma* also this was the characteristic response observed by Verworn. With certain other leaves, again, under fairly strong current, the polar excitations took place at the make of kathode, at the break of kathode, and at the make of anode. This again was like the responsive reaction in *Actinosphaerium* observed by Kühne. I found a very large number of leaves which gave these specific reactions, which precluded the idea of their being accidental. These were most frequently exhibited late in the season and also in winter.

I made many experiments with a view to solve these anomalies, and found that the normal polar reactions of plants are modifiable, to a greater or less extent, by various physiological factors. Among the most important of these may be mentioned the influence of age and of season. Experiments on the effect of age in the modification of response will first be described.

THE EFFECT OF AGE

It is impossible to dissociate from the consideration of the age of a given leaf its past history as regards the stimulus of sunlight. If we examine a plant we find that the youngest leaf is quite green, not having as yet been long exposed to the action of light. The next below it will be older, and owing to the longer action of the sunlight, reddish-brown in colour. The leaves lower down will be older again and still more russet in tint. In this way the sequence of the leaves in point of age, from above downwards, corresponds to the other sequence of duration of exposure to stimulus of light.

It will be remembered that we may broadly classify the condition of a tissue under three different phases: first, the pre-optimum or sub-tonic state; second, the optimum; and third, the post-optimum condition, in which we see an approach towards the condition of fatigue. Moderate stimulation, as we have seen, will carry a tissue out of the pre-optimum towards the optimum condition, with consequent enhancement of excitability. Excessive or too-prolonged stimulation, on the other hand, will carry it to the post-optimum condition with the characteristic depression of its excitability.

From this point of view alone, then, we might expect that the uppermost or youngest leaf of *Mimosa* would be in the pre-optimum and therefore less sensitive condition; that the sensitiveness of the leaves should attain a maximum as we descend lower in the plant; and that after this has been reached, continuing to descend, the excitability of the different leaves will be progressively decreased. Representing these gradations by means of a curve, there would be at first an ascent, then a climax, and after this a sharp turn and descent. Independent of or concomitant with this will be the changes of excitability, more or less obscure, which are brought about by increasing age. Before subjecting this inference to the test, I will describe an experiment which shows that the after-effect of moderate stimulation on a sub-tonic tissue is to raise its susceptibility to polar excitation.

After-Effect of Moderate Stimulation

For this purpose I used a specimen which appeared to be in a pre-optimum condition, and took its records under

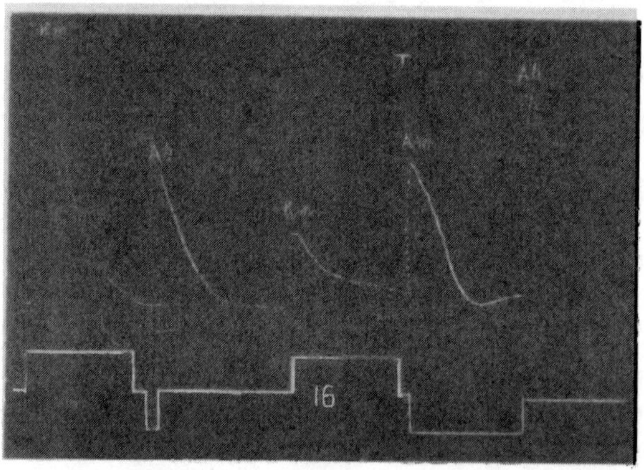

Fig. 127.—Transformation of Type II. to Type III. as after-effect of previous stimulation. First two responses taken before, and last three after, tetanisation.

polar excitation by an identical current, before and after the application of tetanising shocks of moderate intensity and duration.

Under an E.M.F. of 16 volts, the records obtained show excitation (fig. 127) at kathode-make and anode-break, characteristic of Type II. The record was then stopped and tetanising electric shocks of moderate intensity applied for 2 minutes. After a period of rest of 15 minutes, the record of polar excitation T was taken once more, the applied

E.M.F. being the same as before—namely, 16 volts. It will be remembered that before tetanisation 16 volts, though effective for Ab excitation, had been ineffective for Am excitation. After tetanisation, however, the Am excitation became effective. The after-effect of moderate stimulation had thus been to transform **Km Ab**, the polar reaction of Type II., to **Km Am Ab** of Type III.

As already mentioned, from the point of view of stimulus the youngest green leaf at the top of the plant, say L1, may be regarded as being in a sub-tonic condition. A leaf lower down, L2, must be taken as having been previously

TABLE IV.—SIMULTANEOUS ACTION OF CURRENTS OF INCREASING INTENSITY ON A VERY YOUNG L1, AND A SLIGHTLY OLDER LEAF L2

Effective intensity of current.	Resulting types of reaction.			
	I. **Km**	II. **Km Ab**	III. **Km Am Ab**	IV. **Km Kb Am Ab**
	micro-amperes	micro-amperes	micro-amperes	micro-amperes
For L1 ..	4 to 8	12·7	20	—
For L2 ..	4 to 5	5·6	6·3	20

subjected to moderate stimulation. From the experiment just described, then, we should expect that one identical current would evoke response of a higher type from L2 than from L1.

In order to test this inference I took a plant and made bi-polar connections with the second and fourth leaves from the top, the former of these being very young. Besides these, a second pair of bi-polar connections was also made, between the fifth and seventh leaves in descent, and these will be referred to as L3, which was old, and L4 which was very old. The results obtained from the second pair will be dealt with separately.

Above I give a tabular statement of the results obtained

with L1 and L2. The mode of procedure was to apply a current which was gradually increased till successive types of responsive reactions were observed in one or other of the two leaves :—

It is apparent that while a maximum current of 8 micro-amperes induced responses of Type I. in the very young leaf L1, a very much feebler current, of 5·6 micro-amperes was sufficient in the case of L2 to induce Type II. A similar relative exaltation of effect in slightly older specimens occurs in subsequent types also. Thus while 12·7 micro-amperes gives rise in L1 to Type II., half that intensity is enough to induce Type III. in L2. And lastly, while 20 micro-amperes gives Type III. in L1, the same current gives Type IV. in L2.

Having thus verified our inference with regard to the relative excitabilities of two young leaves, we may expect contrasted results with older leaves, where the excitability will be on the wane. Table V. shows the comparative effects of increasing currents on leaves of the same plant, L3 and L4, which are old and very old respectively. For greater facility of comparison the effects on L1 and L2 are repeated.

TABLE V.—POLAR EFFECTS OF CURRENTS ON LEAVES OF DIFFERENT AGE

	I. **Km**	II. **Km Ab**	III. **Km Am Ab**	IV. **Km Kb Am Ab**
	micro-amperes	micro-amperes	micro-amperes	micro-amperes
L1	4 to 8	12·7	20	—
L2	4 to 5	5·6	6·3	20
L3	4 to 16	—	20	23
L4	4 to 21	—	23	—

From this experiment it will be seen, (1) that the general excitability, having reached a maximum in L2, has undergone a progressive decline with age; (2) that while the excitatory efficiency of Km has undergone but slight decline, that of Ab is very marked. This is seen in the fact that while Ab excitation was exhibited by the youngest leaf L1 under 12·7 micro-amperes, and by L2 under 5·6 micro-amperes, no such excitation was exhibited by L3 and L4 under a current as strong as 19 micro-amperes. The decline in excitatory efficiency of Am, on the other hand, was but slight ; in L1 anode-make excitation was induced by 20 micro-amperes, in L2 by 6·3 micro-amperes, in L3 by 20 micro-amperes, and in L4 by 23 micro-amperes. It will be noticed that in the two leaves L3 and L4 there was no excitation of type **Km Ab**.

Apparent Vanishing of Type II, **Km Ab**

We will now consider the possibility of the appearance of new types of polar reaction by the progressive diminution of Ab excitation. A reference to Table V. will show that by

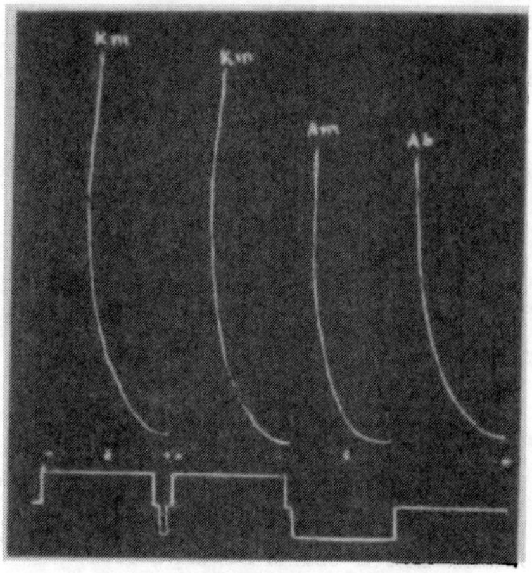

Fig. 128.—Abrupt transition from Type I. to Type III., with the vanishing of Type II.

its greater loss of excitatory efficiency Ab has been reduced to the level of the excitatory reaction of Am. Normally speaking, the excitation induced by anode-break Ab takes place earlier than that caused by anode-make Am. And this is the reason why **Km Ab**, or Type II., precedes **Km Am Ab**, or Type III. But now when **Am** is equal to **Ab** the latter cannot precede, and the two occur at the same minimal intensity of current. Hence the sequence of excitation with increasing current is I., **Km** ; II., absent ; and III., **Km Am Ab**. A series of records are shown in fig. 128, in which Type I., **Km**., passes abruptly to Type III., **Km Am Ab**, as the acting E.M.F. is slightly increased from 5 to 6 volts. It may be stated here that late in the season, and with somewhat old leaves, I was frequently puzzled by this vanishing of Type II. ; but the quantitative explanation which has just been given will adequately account for this.

We have seen that under certain conditions the Ab excitation declines at a greater rate than that of Am, and have considered the resulting effect when the two become equal. In certain circumstances it is quite conceivable that this relative decline of susceptibility to excitation by anode-break might proceed further. It may then happen that Am becomes more effective than

TABLE VI.—MODIFICATIONS OF NORMAL TYPES BY DECLINE OF AB IN EXCITATORY EFFICIENCY

Normal types Ab > Am	Modification (a) Ab = Am	Modification (b) Ab < Am
I., **Km** II., **Km Ab** III., **Km Am Ab**	I., **Km** II., **O** III., **Km Am Ab**	I., **Km** II′., **Km Am** III., **Km Am Ab**

Ab. The sequences of excitatory effects would then be I., **Km ;** II′., **Km Am ;** III., **Km Am Ab.** It will here be noticed that to avoid adding to the number of types, I have designated **Km Am**—which takes the place of **Km Ab**—as II′. or transitional II. In the accompanying table these theoretical modifications of type, due to the decline in excitatory efficiency of Ab, will be seen displayed in a convenient form. Under normal conditions the excitatory efficiency of Ab is greater than Am ; under modification it may become (a) equal to, or (b) less efficient than, Am.

MODIFIED RESPONSE **Km Am**

The existence of modification (a) with the vanishing of **Km Ab** has already been demonstrated. We next come to the theoretical possibility of a modified type **Km Am.** As already stated, in certain circumstances, especially late

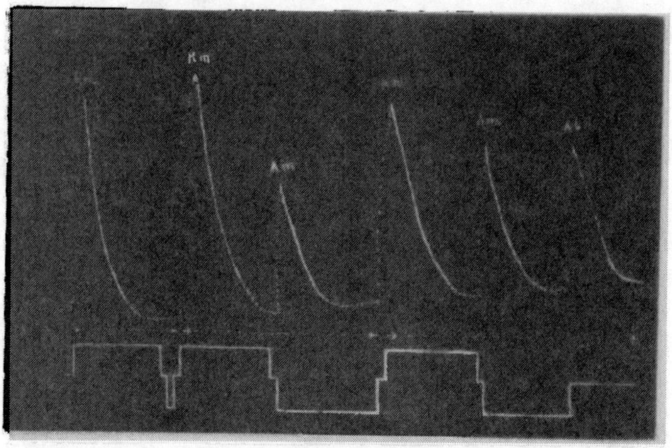

FIG. 129—Polar reactions, **Km, Km Am, Km Am Ab,** under gradually increasing current.

in the season and in older specimens, I have not infrequently met with this type. A series of records are reproduced in fig. 129, in three cycles of effects of increasing current. In these the acting E.M.F. was increased from 6 to 8 volts and then to 10 volts. It will be seen that the stage **Km Ab** has apparently vanished, its place being taken by **Km Am.** An account of results obtained with four different specimens, under gradually increasing E.M.F., is given in the following table, the reaction under a particular type being indicated by an inclined cross.

It is clear therefore that we have in all these cases a

vanishing of Type **Km Ab** and a substitution of the transitional type **Km Am.** It may be mentioned here that in some cases this characteristic excitation, at make only of both kathode and anode, was found to persist through an extended range of current-intensities. The resemblance of this type to the response of *Amphistigma* is obvious.

TABLE VII.—EXPERIMENTAL RESULTS WITH DIFFERENT SPECIMENS EXHIBITING **Km Am** EFFECT

Specimen	E.M.F.	Km	Km Ab	Km Am	Km Am Ab
	volts				
1 ..	6	X	—	—	—
	8	—	—	X	—
	10	—	—	—	X
2 ..	4	X	—	—	—
	6	—	—	X	—
3 ..	2	X	—	—	—
	4	—	—	X	—
	6	—	—	—	X
4 ..	12	X	—	—	—
	18	—	—	X	—
	22	—	—	—	X

MODIFIED RESPONSE **Km Kb Am**

If the excitatory efficiency of Ab were to fall still further, it might be reduced even lower than the excitatory effectiveness of kathode-break. In such a case, with a given current, excitation would be less for Ab than for Am or Kb. Hence the anode-break effect would vanish and the excitatory formula for this type under a certain intensity of current would be **Km Kb Am,** and this we may call III., *Transitional.* It is interesting to find that this particular responsive modification is not seldom obtained, especially with leaves the petioles of which are very thin.

The following table shows such results obtained from four different specimens :—

TABLE VIII.—EXHIBITION OF **Km Kb Am** EFFECT

Specimen	E.M.F.	**Km**	**Km Ab**	**Km Am**	**Km Am Ab**	**Km Kb Ab**
	volts					
1 ..	10	X	—	—	—	—
	12	—	—	X	—	—
	16	—	—	—	—	X
2 ..	2	X	—	—	—	—
	4	—	—	X	—	—
	6 to 10	—	—	—	—	X
3 ..	4 to 9	X	—	—	—	—
	12 to 18	—	—	X	—	—
	20	—	—	—	—	X
4 ..	4	X	—	—	—	—
	16	—	—	—	—	X

A record which exhibits the **Km Kb Am** excitation is given in fig. 130.

We may conclude by briefly reviewing the polar reactions

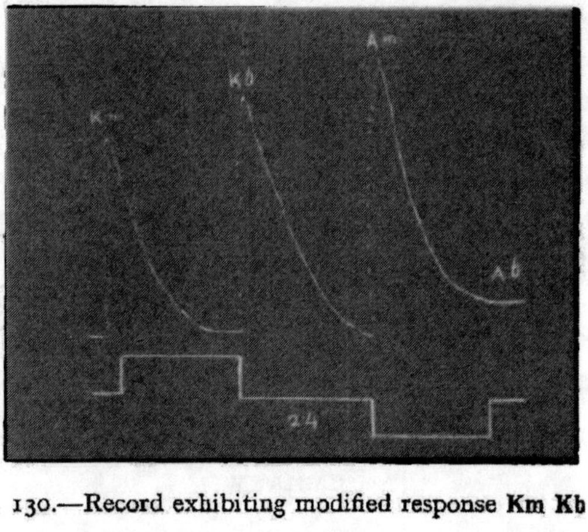

FIG. 130.—Record exhibiting modified response **Km Kb Am.**

of living tissues with the theories and inferences that are current on the subject.

In the first place there are many who hold the universality

of Pfluger's Law. Against their views is brought the fact that in the unfibrillated protoplasm of *Protozoa* the reactions observed are more or less opposed to those of Pfluger's generalisation.

Next are those who hold with Verworn that, in view of the anomalous reactions of unfibrillated protoplasm in *Protozoa*, there could not possibly be any law of polar reactions of universal applicability.

There is, however, a third view still possible. I have shown that the reactions of fibrillated and unfibrillated protoplasm are not necessarily opposed, since the plant, within certain limiting values of current, shows reactions identical with those which are normal in animal tissues.

If then we find that the generally recognised reactions of stage I.—namely, response at kathode-make only—can be transformed into those of stage II.—response at kathode-make and anode-break—by merely increasing the intensity of current, then there can be no inherent impossibility in further transformations into Types III. and IV. under still further increase of the current-intensity.

The possibility of such transformations has now I think been demonstrated by the experiments described in this and previous chapters on numerous species of sensitive plants. It has also been shown, further, that these polar reactions were subject to variation under physiological modification of the tissues. And finally, in plants whose reactions were already established as identical with those of animal tissues, it has been shown that—under the separate or joint action of an increasing current and physiological modification—responses could be obtained which were similar to the so-called anomalous responses characteristic of *Protozoa*.

Summary

The after-effect of moderate stimulation converts the polar reaction of a given type to one of higher type.

Under the same exciting current the resulting type of

reaction depends on the age of the specimen. A moderately young specimen gives higher type of reaction than one which is very young or one which is old.

On account of modification of the tissue, its excitability to anode-break undergoes progressive diminution. In consequence of this, modified types of reaction **Km Am** and **Km Kb Am** are sometimes found to occur in *Mimosa.*

The polar reactions in *Protozoa* are not exceptional. Similar effects are observed in *Mimosa* under specific conditions.

CHAPTER XX

MULTIPLE AND AUTOMATIC RESPONSE

The Oscillating Recorder—Latent period of *Biophytum*—Refractory period—Response on 'all-or-none' principle—Multiple electrical response to a single strong stimulus—Multiple mechanical response to strong stimulus in *Biophytum* and *Averrhoa*—Continuity of multiple and automatic response—Ordinarily responding *Biophytum* converted into automatically responding condition by excess of stored energy—Automatically responding *Desmodium* converted to ordinarily responding condition by depletion of stored energy.

We have studied in detail the responsive characteristic of the leaf of *Mimosa*. We next take up the study of the responses of the leaflet of *Biophytum*. In doing this, we shall observe a certain new class of phenomena of great theoretical importance come into play.

The difficulties encountered here, however, in the taking of automatic records are extremely great. The leaflets are very slender, and the pull exerted in the course of the excitatory fall is very slight. I overcame the difficulties in the taking of the record by the use of the Optical Lever, in which a moving spot of light either traced the response-curve on a travelling photographic-plate, or was itself followed by the pen of the observer on a moving drum covered with paper. These devices have their disadvantages, and I was desirous of contriving means to secure records at once simple, effective, and perfectly automatic.

The Resonant Recorder was not found specially suitable for the tracing of *Biophytum* movements, inasmuch as the lightest steel wire was still too heavy for the slight pull exerted by the leaflet; the only lever that was sufficiently light to give a slightly magnified record of these movements was some special kind of dry grass haulm, which combined rigidity with excessive lightness. I have tried the finest feathers from small birds, but these were not so efficient

FIG. 131.—Photograph of the Oscillating Recorder, reduced to one-fourth the natural size.

as the selected specimen of grass which I was so fortunate as to obtain later. Though in this I was successful in obtaining a recording-lever, there yet followed the difficulty that owing to its being non-magnetic it could not be thrown into resonant vibration by the electro-magnet. I had therefore to devise some arrangement by which the recording-plate could be maintained in a state of oscillation. The recording-plate was consequently made as light as possible, using the glass plate employed for covering magic-lantern slides; the carrying-frame was made of aluminium. By means of an electric motor and an eccentric device, the plate-carrier was made to oscillate to and fro, the frequency of oscillation being regulated by an adjustment of the electrical current (fig. 131).

The plate-carrier has small wheels which run between horizontal rails, above and below. The recording-plate thus travels in a horizontal instead of vertical direction. There is also a knock-over key or trigger arrangement, not shown in the figure. During a particular part of the travel of the plate the trigger is released, causing a single break-induction-shock to pass through the plant. The moment of stimulation is marked in the usual manner on the travelling-plate, and the number of dots intervening between this mark and the beginning of response enables us to determine the latent period when the stimulation is direct, or the velocity of transmission when the stimulation is indirect. In the latter case it is of course necessary to know the distance between the point of application of stimulus on the midrib and the responding leaflet.

For the determination of the latent period, and of the velocity of transmission, the oscillation-frequency of the plate should be about 10 times in a second. But for the mere obtaining of response-records the frequency of oscillation need not be high.

Determination of the Latent Period

Taking a record of *Biophytum* on a fast-moving plate, with a recorder having a frequency of oscillation of 10 times a second, it is seen (fig. 132) that there are four spaces between the incidence of stimulus and the initiation of response; it would thus appear that the latent period of the leaflet is ·4 of a second. This seems very high compared with the latent period of *Mimosa*, which is ·1 second. Perhaps this difference may be due to certain characteristics that mark the response of *Biophytum*. In animal tissues it is found that while the singly responding skeletal tissue of the frog has a latent period of about ·01 second, its multiple responding cardiac tissue has a latent period of about ·1 second, or ten times as long. We shall presently see that the leaflet of *Biophytum* exhibits multiple response.

If a second stimulus be applied a short time after the first, it is found that it is ineffective unless a certain minimum

FIG. 132.—Record giving the latent period of the leaflet of *Biophytum*. Frequency of vibration of recorder 10 times per second.

interval of time elapses between the two. In these circumstances the leaflet takes no account of the second stimulus, becoming apparently refractory to it. The minimum interval that must elapse before the second stimulus can be effective—the refractory period—varies somewhat in different specimens. In *Biophytum* it is usually 10 seconds.

Response on All-or-none Principle

In a previous chapter I have given the record of a response in *Biophytum* under a single stimulus (fig. 15). We will now study the effect of increasing intensities of stimulus on the amplitude of response. We have seen that in the case of *Mimosa* increasing intensities of stimulus induce, generally speaking, increasing amplitudes of response, which however reach a limit. The same is true of the responses of a skeletal muscle of frog.

In carrying out experiments on *Biophytum*, I first determined the minimal intensity of induction-shock that was effective in inducing response. This happened when the intensity of stimulus was ·1 unit. The record of this response under minimal stimulation was then taken. After this, a second response to stimulus which was ten times as strong, was recorded. It will be seen that both minimal and maximal stimuli induced practically the same effect (fig. 133). In other words, we have here an example of what is

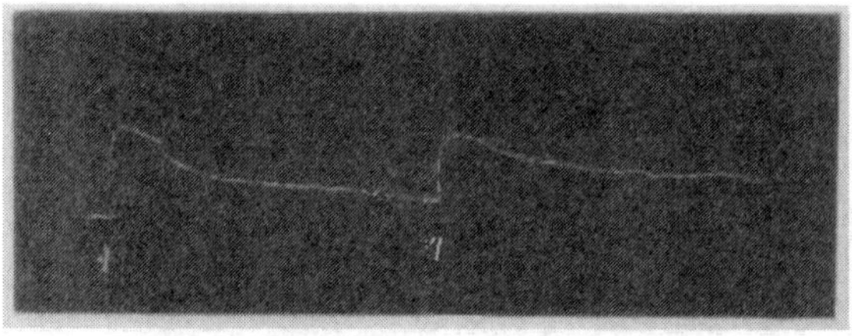

FIG. 133.—Record of responses of *Biophytum* to stimuli ·1 and 1 unit, respectively.

known as response on the 'all-or-none' principle. The leaflet either responds to its fullest or not at all.

In the various characteristics which have just been described, the responses of the vegetal organism bear a curious resemblance to those of the cardiac tissue of the animal. In the response of the animal heart the latent period is relatively long, and it exhibits a similar prolonged refractory period. Its responses are also on the 'all-or-none' principle—it either responds to the utmost or not at all.

Returning to *Biophytum*, we have seen that in order to induce any response a certain minimal intensity of stimulus was necessary. But when the intensity of this stimulus was further increased, the outward expression, or response of the leaflet, remained apparently as before. What then became of the excess of energy that impinged upon it in the form of stimulus? It is not necessary to suppose that in every instance the whole energy imparted by stimulus gives rise to useful work. Some of it may be wasted as heat. But, on the other hand, it is conceivable that the excess of this energy may be stored up, for the time being, to find subsequent expression. To take a physical illustration, the energy stored up in a compressed spring may, on release, give rise to long-continued and rhythmic oscillations. The question arises, then, whether in the leaflet of *Biophytum*, impressed by an excess of stimulation, there may be any analogous storage of energy. Supposing this to take place, the superfluous energy might be utilised to do some internal work not discernible by the observer; or it might, in favourable circumstances and in the presence of suitable motile indicators, find expression in rhythmic movements.

I have referred elsewhere to another mode of recording excitatory response in plants—namely, the electrical. Using this, I have frequently observed that although under a moderate stimulus a single stimulus gives rise to a single electric response, yet under a strong stimulus there would arise a series of responses. Thus, while a feeble stimulus induced a single response, a strong stimulus gave rise to multiple responses.

Having this in mind and the peculiar characteristics of *Biophytum* response, I expected to demonstrate the occurrence of multiple responses by means of mechanical movements. It had already been noticed that *Biophytum* when strongly excited closed its leaflets, not by one but by two successive twitches. It appeared to me that this curious phenomenon was parallel to the multiple response of rhythmic animal tissues; and I expected, if this were so, that instead of two it would be possible to obtain from it a long series of rhythmic responses comparable with the multiple rhythmic responses in animal tissues. In order to obtain this in the case of *Biophytum* it was necessary to prevent the complete closure of the leaflet, in consequence of which further exhibition of mechanical response is rendered impossible. This was secured by applying a light counterpoise in the second arm of the lever. The effect of this was by exerting a tension to hasten recovery, and thus oppose the complete closure of the leaflet.

I then took records of responses of *Biophytum* to quan-

FIG. 134.—Response of *Biophytum* leaflet to stimuli ·1, ·5, 1, and 2. Response is seen to be multiple with the last.

FIG. 135.—Multiple response in *Biophytum* under a single strong electric shock.

titative stimuli of increasing intensity, the stimuli being applied at intervals of 3 minutes (fig. 134). The successive stimuli were of intensities ·1, ·5, 1, and 2. Owing to incomplete recovery during the intervening resting-intervals, the base-line is seen to be shifted upwards. The amplitude of successive responses is about the same, though the stimuli are increasing. At the application of the fourth

FIG. 136.—Multiple response in *Averrhoa* under a single strong electrical stimulus. Vertical marks below indicate time-interval of 1 minute in this and in the following records.

stimulus, of intensity 2, we find that the response becomes multiple. Thus we see that while a single stimulus of moderate intensity gives rise to a single response, a strong

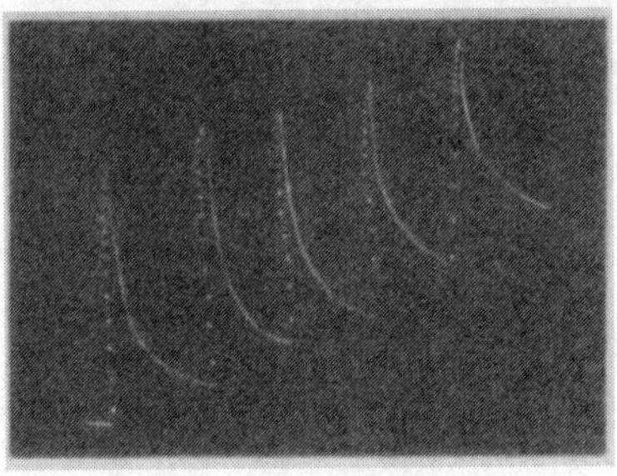

FIG. 137.—Multiple excitation in *Biophytum* under the action of constant light.

stimulus gives rise to a multiple series of responses. In fig. 135 are depicted multiple responses in *Biophytum* to a very strong electrical stimulus, there being four multiple responses in the course of 6 minutes. In fig. 136 are shown the multiple responses obtained in *Averrhoa carambola* under a single stimulation caused by strong induction-shock. Here there are six responses in the course of 15 minutes, the average period of a single pulsation being 2·5 minutes.

We have also seen multiple responses induced in *Biophytum* under the action of constant current (cf. fig. 124). Multiple responses also take place in *Biophytum* and in *Averrhoa* under the action of constant light. I give a

FIG. 138.—Multiple response under the action of single strong thermal shock.

record (fig. 137) in which is seen the occurrence of five multiple-responses in the course of 8 minutes under the continued action of light from an arc-lamp.

I have also obtained multiple responses under other forms of strong stimulation. In fig. 138 is given a record of multiple responses induced by the action of a strong thermal stimulus. In certain other cases I obtained with *Biophytum* as many as sixteen recurrent responses under a single thermal shock.

In fig. 139 is seen a series of multiple responses induced by strong chemical stimulation. This was caused by the application of a drop of hydrochloric acid on the petiole.

From these experiments it is clear that a rhythmic series of effects need not have a periodic antecedent cause. We see on the other hand that under strong stimulation there is not only an immediate response but that the surplus of energy remains over and is held latent by the tissue to be given out later in the form of recurrent responses.

It is the excess of latent or *Internal Energy* that gives rise to phenomenon of multiple response. Sometimes we may not have noticed the antecedent external stimuli the

FIG. 139.—Multiple response induced by strong chemical stimulation.

absorption of which had contributed to that storage of internal energy which gave rise to the rhythmic activity. In these circumstances the pulsations appear to us as spontaneous or automatic.

Under natural conditions, the plant is exposed to the action of various stimuli supplied by its environment. It is exposed to warmth, to the action of light, to internal hydrostatic pressure, to the action of various chemical agents—present in it or absorbed by it. We have seen that each of these factors exerts its stimulating action independently. From the joint action of these external sources of stimulation, the energy stored up by the plant may become sufficiently great to cause an excitatory overflow. It will thus be seen how, by the cumulative effects of these various stimuli, the excitability of the plant may become so excessive as to manifest itself by outward response, apparently automatic.

Thus I have obtained from *Biophytum* seemingly automatic pulsation by subjecting a vigorous plant to the favourable conditions of light and warmth. In a particular case the favourable temperature was found to be 35° C. When this was lowered to 29° C. the pulsations came to a stop.

It is thus clearly seen that there is no strict line of demarcation between multiple and automatic responses so called. An ordinarily responding plant like *Biophytum*, which gives a single response to a single moderate stimulus, and multiple responses to a strong stimulus, will in very favourable circumstances, that is to say, when it has absorbed an excess of energy from without, become automatically responding.

Conversely, an automatically responding plant in unfavourable circumstances is found to be converted to the condition of an ordinarily responding plant. The leaflet of *Desmodium gyrans* under favourable conditions is found to execute pulsatory movements which appear to be spontaneous. But when this plant is subjected to unfavourable conditions, then its spontaneous rhythmic activity comes to a stop. In this condition of standstill, the reaction of the leaflet is like that of *Biophytum*. It then gives rise to a single response to a single moderate stimulus, and multiple responses to a strong stimulus.

There is thus seen a continuity in the multiple and automatic responses. *Biophytum* is equivalent to *Desmodium* when brought to a state of standstill by depletion of storage of energy. Pulsating leaflets of *Desmodium* may, on the other hand, be regarded as equivalent to *Biophytum* with an overflow of energy.

Summary

In plant a single moderate stimulus is found to give rise to a single electric response. A strong stimulus, on the other hand, often gives rise to a multiple series of electric responses.

In *Biophytum*, similarly, while a moderate stimulus gives rise to a single mechanical response, a strong stimulus gives rise to a multiple series of responses.

These multiple responses are induced by various modes of strong stimulation such as induction-shock, constant current, strong light, thermal shock, and chemical excitation.

Certain plant tissues have the power of holding the excess of stimulus latent, to be given out later in the form of recurrent responses.

The characteristics of the response of *Biophytum* are like those of cardiac tissue of the animal. Both are characterised by long refractory period and response on 'all-or-none' principle. A single moderate stimulus gives rise to a single response in both, and a strong stimulus gives rise to a multiple series of responses.

There is no strict line of demarcation between the phenomena of multiple and of automatic response. In very favourable circumstances of absorption of excess of energy from without, an ordinarily responding plant like *Biophytum* will become converted into an apparently automatically responding plant like *Desmodium*.

Conversely, under unfavourable conditions, that is to say, when the sum-total of its energy is below par, an automatically responding plant like *Desmodium* will become converted into an ordinarily responding plant like *Biophytum*. Its leaflets then come to a state of standstill.

CHAPTER XXI

THE AUTOMATIC PULSATIONS OF *DESMODIUM GYRANS*

Activity of detached leaflet of *Desmodium*—Pulsation maintained uniform under constant internal hydrostatic pressure—The plant chamber—Time-relations of pulsating movement derived from dotted record—Significance of down and up movements—Systole and diastole—Table showing rates of movement of *Desmodium* leaflet at different phases.

WE have hitherto studied the responsive movement in sensitive plants, where such movement was initiated by a directly exciting stimulus. We shall now take up the consideration of another class of movements, which are apparently automatic or without any immediately preceding cause. In certain plants we observe what are known as spontaneous movements, of which *Desmodium gyrans* or the telegraph-plant furnishes an example.

FIG. 140.—Leaf of *Desmodium gyrans*. The two small lateral leaflets exhibit spontaneous movements.

This telegraph-plant grows wild in the Gangetic plain; its Indian name is *Bon Charal*, or 'forest churl,' the popular belief being that it dances to the clapping of the hands. There is, however, no foundation for this belief. It is a papilionaceous plant with trifoliate leaves, of which the terminal leaflet is large, and the two lateral, very small (fig. 140). Each of these is inserted on the petiole by means of pulvinule. The lateral leaflets are seen to execute pulsating movements which are apparently uncaused, and are not unlike the rhythmic movement of the heart. The extraordinary similarity under various conditions of the rhythmic reactions in the plant and the animal will be seen in the experiments which will be presently described.

It will, moreover, be shown that, strictly speaking, there is no such thing as spontaneous movement. The energy which expresses itself in pulsating activity is derived by the plant either directly from immediate external sources or from the excess of such energy already accumulated and held latent in the tissue. When the storage is exhausted, the spontaneous movement, so called, is found to cease. In this condition of standstill the rhythmic activity can be renewed by an accession of fresh stimulus.

In the intact plant, under favourable conditions, these spontaneous movements are observed to take place more or less continuously; but there are times when they come to a standstill. For this reason and because of the fact that a large plant cannot easily be manipulated as a whole and subjected to the various changing conditions which the purposes of investigation demand, it is desirable if possible to experiment with the detached petiole carrying the pulsating leaflet. The required amputation, however, may be followed by arrest of the pulsating movements. But, as in the case of the isolated heart in a state of standstill, I find that the movement of the leaflet can be renewed in the detached specimen by the application of internal hydrostatic pressure. Under these conditions, the rhythmic pulsations are easily maintained uniform for many hours. This is a great advantage, inasmuch as in the undetached specimens the pulsations are not usually found to be so regular as they now become. So small a specimen, again, can easily be subjected to changing experimental conditions, such as variations of internal hydrostatic pressure and temperature, application of different drugs, vapours, and gases.

The petiole after detachment should be put in water immediately, to prevent complications arising from drying of the cut end. It is then mounted water-tight, in the shorter open end of a narrow U-tube filled with water. For this purpose an indiarubber cork is taken, and a slit made from circumference to centre. The petiole is then slipped into the centre of the cork, which is gently forced on to the short open end of the U-tube. The longer end of the U-tube partly consists of indiarubber tubing. By raising or lower-

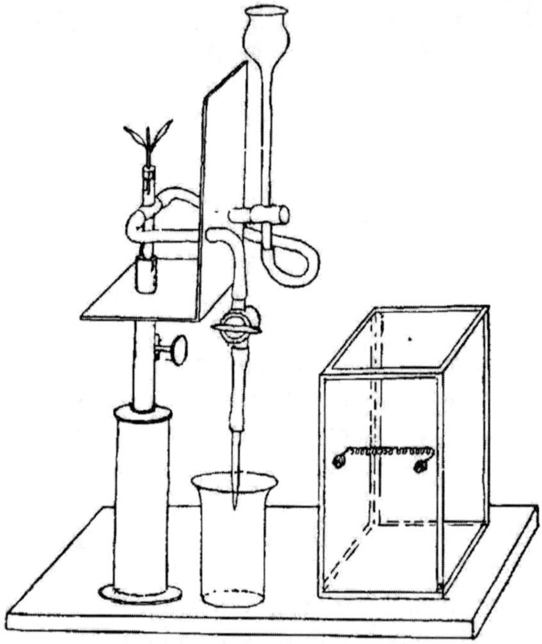

FIG. 141.—U-tube support for leaflet, and the plant chamber.

ing this longer limb of the U-tube the hydrostatic pressure to which the specimen is being subjected can be varied; different chemical solutions can also be applied internally by its means; a stopcock allows the water to run out of the U-tube, making way for the particular solution poured in at the open end of the tube.

The U-tube (fig. 141) is hinged on a rod which slides up and down inside an upright tubular support. This rod can also be rotated inside its support and clamped in any desired position. Facilities are thus secured for three different modes of adjustment. One up and down, the second lateral, and the third, by means of the hinged support, for the inclining of the specimen. The movement of the leaflets, it must be remembered, does not always take place in a vertical plane. The object of these mechanical adjustments therefore is to enable us to place the specimen at such an angle that its to-and-fro vibration when straight shall be vertical, or have its long axis vertical when the movement is elliptical.

A light cover with mica windows can be made to enclose the specimen. By means of electric current sent through a spiral of German silver, the inside of the chamber may be heated to any desired degree. The temperature can be lowered, on the other hand, by sending through the chamber a stream of cooled air: different vapours or gases could be passed into the chamber for studying their effects on the automatic pulsation.

The arm of the recording-lever is attached by means of a cocoon thread to a point about the middle of the leaflet, by means of a drop of shellac-varnish. As the pull exerted by these leaflets is very feeble, the writer has to be made extremely light. The vibration of the Resonant Recorder being about 10 times per second, the record taken with it appears as continuous. In certain experiments it is desirable to obtain data for accurate time-measurements of different phasic movements of the leaflet. This I have been able to secure by the employment of the Oscillating Recorder, where the recording-plate, by means of an electric motor provided with an eccentric, is made to execute a to-and-fro movement. The intermittent dots thus produced may be once each second, or once in 2 seconds. As the oscillating recorder permits the employment of light grass haulm for the recorder, we may easily obtain a fair magnification produced in the record. I have used both the methods—resonant and oscillating—for obtaining the records: in the former they appear continuous; in the latter, dotted.

As an example of the extreme regularity which can be

secured in the pulsating movement of the cut specimens, I reproduce here a continuous record lasting for four hours (fig. 142), the movements themselves being maintained uniform for more than seven hours. The run of the breadth of the plate was accomplished in one hour and twenty minutes, successive series of records being taken on the same plate from below to above. It will be seen how uniform are the successive pulsations, not only as regards

FIG. 142.—Continuous record of pulsations of *Desmodium* leaflet for four hours; the series to be read from below to above.

the amplitude, but also the period. It is only after securing such uniformity, under normal standard conditions, that the experimenter is justified in drawing correct inferences, from variations induced in the record, on the influence of changed conditions which he has introduced.

TIME-RELATIONS OF THE PULSATING MOVEMENTS OF *DESMODIUM* LEAFLET

The up-and-down movement of the *Desmodium* leaflet is characterised by certain peculiarities. In its usual form the 'up' movement is executed somewhat slowly, and comes to what may be called a pause at the extreme position for a certain length of time. There is in reality no cessation of movement; but the rate becomes extremely slow at the turning-point where 'up' is reversed to 'down' movement and *vice versa.* After reaching the extreme 'up' position the 'down' movement is commenced, and this is accomplished in a much shorter time. After reaching the lowest position there is again a pause, when the cycle is again repeated. These normal movements and their rates are, however, subject to modification under the influence of external conditions.

These movements of *Desmodium* leaflet are brought about, as in the case of *Mimosa*, by the contraction or expansion of the pulvinus. Here, also, it is the lower half of the motile organ that is predominant in its action. A question now arises as to the significance, whether of contraction or relaxation, of the up and down movements. The question may be settled in three different ways: The contractile movement, generally speaking, is quicker; hence the quicker down movement of the leaflet may be regarded as that due to contraction. Again, I have found that a leaflet in a state of standstill exhibits under stimulation an excitatory contractile movement which is downwards. And lastly, by means of internal hydrostatic pressure, we may induce expansion of tissue. This has the effect, as we shall see in a later chapter, of shifting the pulsatory movements upwards. All these different considerations point to the conclusion that in *Desmodium* the 'down' position of the leaflet represents contraction and the 'up' position denotes expansion of the more effective lower half of the motile organ. The up-and-down movements of the leaflet thus correspond to the diastolic and systolic movements of the animal heart. In the records of the pulsation of *Desmodium*, up-curve represents down movement and *vice versa.*

It would be interesting to know the absolute rate of

these movements in their various phases. The movements themselves are relatively slow, much slower than the contractile movement of *Mimosa*. In that case we saw that the movement of the fall was accomplished in about 2 seconds but with *Desmodium* the down movement may occupy as long as 40 seconds. I was able to determine the different rates of movement in *Desmodium* by making the recording

FIG. 143.—Record of a single pulsation of *Desmodium*; magnification 2·5 times. Successive dots at intervals of a second.

plate oscillate to and fro once in a second or once in two seconds.

In fig. 143 is given a record of a single pulsation, magnified 2·5 times and taken on a fast-moving plate, the successive dots being at intervals of a second. The period of an entire pulsation was 101 seconds, of which the down movement was accomplished in 41 seconds and the up movement in 60 seconds. The spacing of the successive dots at once gives a visual representation of the changing rate. It is noticed that the leaflet attains its maximum rate during the fourteenth second of its downward journey. The maximum rate of the down-movement is ·9 mm., the average rate being ·44 mm. per second. The maximum rate of the up movement was ·56, and the average rate ·3 mm. per second.

In fig. 144 is shown a record of two successive pulsations obtained with a different specimen. The amplitude is

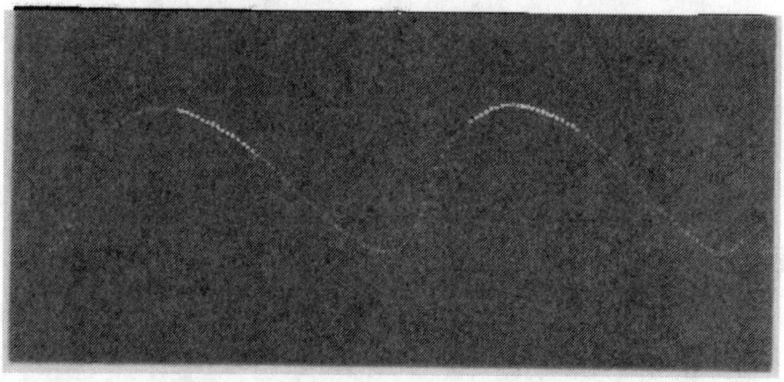

FIG. 144.—Records of two pulsations of a different specimen of *Desmodium* reduced to ⅔. Successive dots are at intervals of 2 seconds.

reduced to two-thirds in the diagram and the successive dots are at intervals of 2 seconds. A detailed account is given in the accompanying table of the rates of movement exhibited by the two specimens :—

TABLE SHOWING RATES OF MOVEMENT AT DIFFERENT PHASES OF PULSATION OF *DESMODIUM*

——	Specimen 1	Specimen 2
Down movement		
Total period	41 seconds	42 seconds
Average rate	·44 mm. per sec.	·46 mm. per sec.
Maximum rate ..	·9 mm. per sec.	1·0 mm. per sec.
Up movement		
Total period	60 seconds	92 seconds
Average rate	·3 mm. per sec.	·21 mm. per sec.
Maximum rate ..	·56 mm. per sec.	·5 mm. per sec.

I give another series of records obtained with a vigorous

specimen taken on a slower-moving plate (fig. 145). This record shows the extreme uniformity of these pulsations. How uniform the time-relations are will appear from the tabular statement below, where the intervals of time are

FIG. 145.—Series of automatic pulsations of *Desmodium*. Successive dots at intervals of 2 seconds.

measured between the attainment of highest and lowest positions :—

TABLE SHOWING PERIODS OF DOWN AND UP MOVEMENT IN *DESMODIUM* PULSATION

No.	Period of down movement	Period of up movement	Total period
	seconds	seconds	seconds
1	32	56	88
2	32	56	88
3	34	54	88
4	36	54	90
5	32	54	86
6	32	54	86

These figures are typical of the movement of *Desmodium* leaflet under normal conditions in the summer season. In winter the rate is very much slower, the period of an entire pulse being then as long as 4 minutes.

Various other factors modify either the amplitude or frequency of the pulsatory movement. One factor which induces a pronounced change is that of temperature. The detailed consideration of this and of other factors will be given at some length in the following chapters.

SUMMARY

The pulsating activity of the detached leaflet of *Desmodium gyrans* can be maintained uniform for a long time by subjecting it to internal hydrostatic pressure.

The application of a shock to a leaflet in a state of standstill induces a down movement. The phase of down movement is in general quicker. Enforced expansion by increased internal hydrostatic pressure induces movement of the leaflet upwards. These facts indicate that the down position of the leaflet represents a 'systolic' contraction, and up position a 'diastolic' relaxation of the motile organ.

In a typical example of the rhythmic pulsation of *Desmodium* leaflet the down movement is accomplished in 41 seconds. The maximum rate of down movement is ·9 mm. per second, the average rate being ·44 mm. per second. The period of up movement is longer, being 60 seconds; maximum rate of up movement is ·56 mm. per second, the average rate being ·3 mm. per second.

CHAPTER XXII

EFFECT OF HYDROSTATIC PRESSURE, LOAD, AND LIGATURE ON THE PULSATION OF *DESMODIUM*

Effect of internal hydrostatic pressure on the pulsation of *Desmodium*——Expansive erection of leaflet under increased pressure : diminution of the extent of systolic contraction—Effect of load : diminution of period—Stannius' ligature on heart-beat—Parallel effect of ligature on pulsation of *Desmodium*—Arrest of pulsation by a cut and revival by electric shock.

WE have already seen that after the application of a suitable internal hydrostatic pressure the pulsation of the detached leaflet of *Desmodium* becomes extremely regular. Sometimes a leaflet is found which has come to a state of standstill. Internal hydrostatic pressure is often found in such cases to renew the pulsatory activity.

EFFECT OF HYDROSTATIC PRESSURE

The application of internal hydrostatic pressure enables us, moreover, to understand the significance of the up-and-down movements. Such movements may be the results of contractions and expansions, as in a pulsating heart. In the case of *Desmodium* we can induce expansion artificially by internal hydrostatic pressure ; the shifting of the base-line up or down will then distinguish for us the result of expansion. Of the up-or-down movement again, the one corresponding to expansion will be helped by increased pressure ; the other, on the contrary, will be opposed.

In subjecting the detached leaflet of *Desmodium* to

increased hydrostatic pressure it is found that the base-line is shifted downwards, which in the leaflet it must be remembered means erection. The contractile effect is also opposed, hence we observe the extent of contraction gradually

FIG. 146.—Application of increased internal hydrostatic pressure at arrow. Displacement of base-line downwards in the record indicates expansive erection of leaflet. Enforced expansion also causes decline in the extent of systolic contraction.

decreased, with the result that a line joining the apices of the successive pulsation slopes downwards (fig. 146).

EFFECT OF LOAD

We will next study the effect of load. Different weights are placed on the second arm of the lever. The effect of the weight is to pull the leaflet slightly upwards. This pull will have the effect of opposing the contractile movement.

I took a series of records of the pulsation of *Desmodium*, first without any weight, then with increasing loads of $\frac{1}{200}$, $\frac{1}{100}$, and $\frac{1}{50}$ grm. (fig. 147). It will be seen that the amplitude of pulsations is in consequence progressively diminished. Without load, the amplitude is 19 mm.; under a load of $\frac{1}{200}$ grm. it has become reduced to 11 mm.;

under $\frac{1}{100}$ grm. to 8 mm., and finally, under $\frac{1}{50}$ grm. to 4 mm. When the load is increased to $\frac{1}{40}$ grm. the pulsation is arrested. The pulsation is also seen to become slower with increasing load.

Arrest of Pulsation by Ligature

A very interesting phenomenon, often observed in a pulsating heart, is the effect of cut or ligature. Thus by the application of what is known as 'Stannius' ligature' above

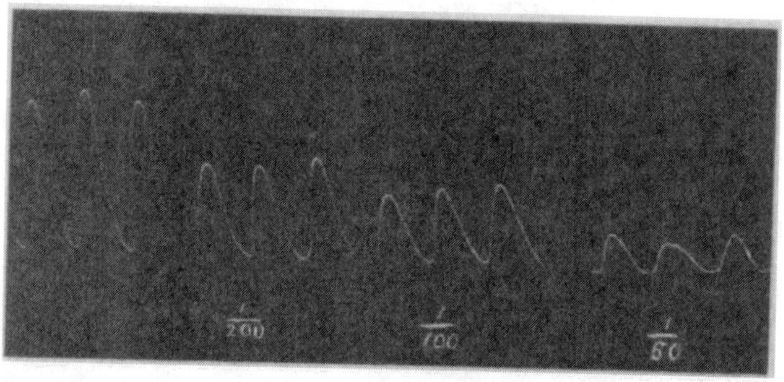

FIG. 147.—Effect of increasing load of $\frac{1}{200}$, $\frac{1}{100}$, and $\frac{1}{50}$ grm. Amplitude of pulsation decreased, and period increased, with increasing load.

the heart, its beat may be arrested. This arrest takes place in a relaxed condition, that is to say, at diastole. The effect, however, is liable to be modified by the condition of the heart. Sometimes there is a failure of arrest. At other times, on the application of ligature there are a few vigorous pulsations followed by arrest. Effects similar to these are also induced by cut. The various explanations hitherto offered of these peculiar effects have all been pronounced untenable for one reason or another.

Seeing the remarkable parallelism which obtains between the pulsation of the cardiac tissue and that of the leaflets of *Desmodium*, I was curious to discover whether in the case of 'Stannius' ligature,' also, there was a similar correspondence. It was a great surprise to me, on applying a cut or a ligature to the petiolule of the leaflet, about 3 mm. below the pulsating pulvinule, to find the various peculiarities recorded of the action of the heart repeated here, with striking similarity.

The most convenient way to apply what was equivalent to the ligature, was to hold the petiolule in a very small clamp and proceed to record the normal pulsation. After doing this for some time, the clamp was suddenly tightened.

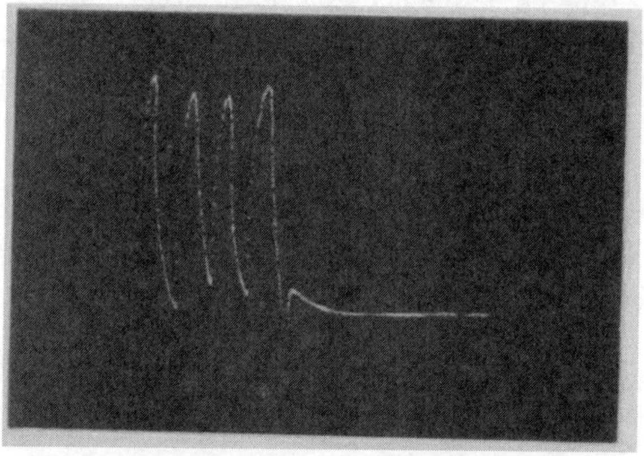

FIG. 148.—Effect of ligature in inducing an arrest of pulsation of *Desmodium* at diastole.

This induced an arrest of pulsation, either at once or after one or two vigorous beats. Similar effects were also obtained by making a cut, after suitably supporting the leaflet. These results depend somewhat on the condition of the specimen. It is generally found that the nearer it is made to the pulvinule the more effective is the cut or ligature in inducing the arrest.

In fig. 148 is seen the arrest induced by ligature : it is very remarkable that here, as in the case of the pulsating heart, the arrest by ligature took place at diastole.

In fig. 149 is seen the arrest induced by a cut. While the leaflet is still in the condition of arrest it is often possible to renew the pulsation by an induction-shock. These phenomena of arrest and revival are easily observable in the record.

The question next arises whether any theory can be suggested as to the cause of this arrest. A possible explanation might lie in regarding the cut or sudden ligature as intense forms of stimulation, whose effect is somewhat persistent rather than instantaneous. We have further to suppose that this intense stimulation is conducted by the intervening petiolule. We have seen, moreover, that the leaf of *Mimosa* under prolonged stimulation, after a preliminary excitatory contractile movement, assumes a

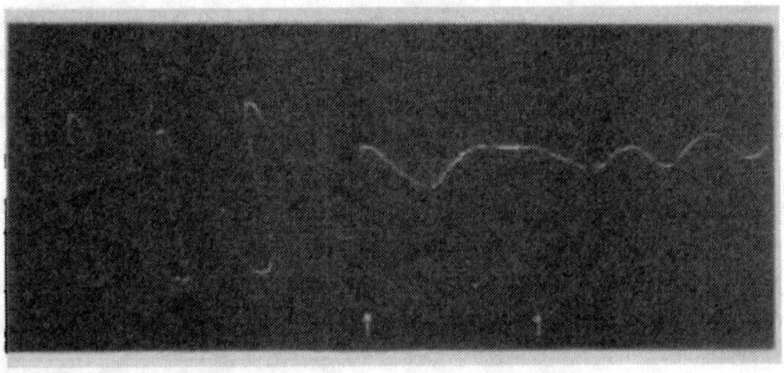

FIG. 149.—Arrest of pulsation of *Desmodium* by a cut applied at moment marked by first arrow. Pulsation was revived by electric shock applied at moment marked by second arrow.

relaxed position indicative of over-stimulation and fatigue. After the lapse of a requisite time, however, the leaf regains its sensitiveness. In the case of the *Desmodium* leaflet a similar effect might conceivably be induced by the transmission of intense excitation from the stimulated cut or ligatured end. This question of the effect of ligature is admittedly one of great difficulty, even in the case of the cardiac tissue. The explanation here offered, as applying to the *Desmodium* leaflet, may be taken as suggestive and more or less tentative. Such an idea presupposes that the intervening petiolule contains some conducting-tissue by which excitation may be transmitted. In any case, the question as to the power of the petiolule to conduct excitation applied at a distance, and thereby induce a modification in the pulsating activity of the attached leaflet, is one of very great importance. This question will be taken up in a succeeding chapter.

SUMMARY

Increased hydrostatic pressure induces an expansive erection of the leaflet of *Desmodium gyrans*. The enforced expansion induces a diminution in the extent of systolic contraction.

Increasing load induces a diminution of amplitude and prolongation of period of pulsation. A load of $\frac{1}{40}$ grm. is enough to arrest the pulsation.

A ligature applied 3 mm. below the motile organ arrests the pulsation of *Desmodium* leaflet at diastole. This corresponds to the action of Stannius' ligature on the heart.

A cut also arrests the pulsation of *Desmodium*. After the arrest, pulsation may sometimes be revived by the action of an electrical shock.

CHAPTER XXIII

EFFECT OF STIMULUS ON LEAFLET OF *DESMODIUM* AT STANDSTILL

Condition of standstill brought about by depletion of energy—Renewal of pulsation by the stimulus of light—Response to stimulus of induction-shock—Multiple response under tetanisation—Determination of the latent period and the apex time—Refractory period—Effect of stimulus on leaflets in sub-tonic condition—Effect of isolation on rhythmic activity—Gradual arrest of pulsation resulting from run-down of stored energy—Effect of fresh accession of energy.

THE leaflets of *Desmodium* exhibit in favourable circumstances a more or less persistent rhythmic activity. But this activity may cease under conditions which are less favourable. The arrest may be due to either of two causes, of which the first is the exhaustion of that reserve of energy without which the pulsation cannot be maintained, and the second is the loss of motility of the pulvinule brought on by age or other factors. As a parallel instance to this we have the case of *Biophytum*, the old leaflets of which become quite insensitive, while the plant as a whole remains sensitive.

RENEWAL OF PULSATING ACTIVITY UNDER THE ACTION OF STIMULUS

In those cases where the arrest is due to run-down of energy, the pulsation of *Desmodium* leaflets may often be renewed by the application of appropriate stimulus. This may be seen in the following record on the action of the stimulus of light (fig. 150). The leaflet had been reduced by extreme sub-tonicity to a state of standstill. The quiescent leaflet was then subjected to the continued action of light from a Nernst lamp. It will be noted that by the absorption of the energy of light the leaflet regained its so-called spontaneous activity.

It should be mentioned here that the effect induced by the stimulus is, to a certain extent, modified by the tonic condition of the plant. In a sub-tonic specimen, with the leaflet in a state of standstill, the action of stimulus is to renew the pulsating activity. In a different specimen,

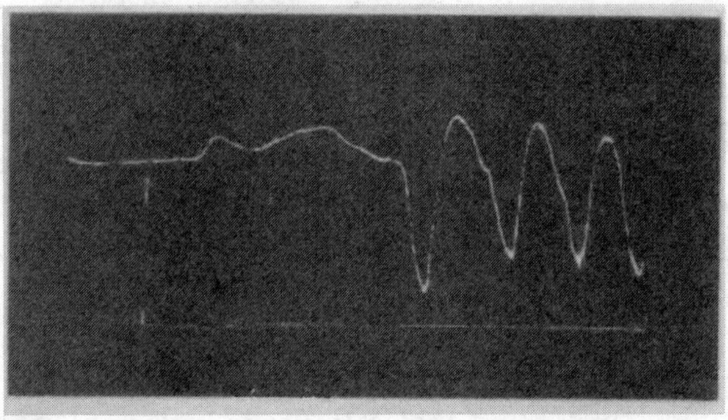

FIG. 150.—Action of light in renewing the pulsation of *Desmodium* leaflet at standstill. Light was applied at the point indicated by the arrow and continued afterwards.

where the rhythmic activity is feeble, the incidence of stimulus enhances the amplitude of pulsation. If the leaflet should be in a vigorous condition, excessive stimulation is apt to bring on fatigue, in consequence of which the pulsations become either irregular or diminished in amplitude. These various effects are found to take place not only under the action of stimulus of light but, as we shall see, under electric stimulus also.

EFFECT OF ELECTRIC STIMULUS

The leaflet of *Desmodium* in a state of standstill may have its activity revived by other modes of stimulation, such as that of induction-shock. This can be seen in the following record (fig. 151). The leaflet had been reduced by extreme sub-tonicity to a state of standstill. A single electrical stimulus of moderate intensity is seen to give rise to a single responsive pulsation. Repetition of the same stimulus gave rise once more to a second response.

FIG. 151.—Response of *Desmodium* leaflet in a state of standstill. Single response to single stimulus.

In another instance, direct tetanising electric shock of short duration was applied to a leaflet in a state of standstill. It is seen (fig. 152) that this application of moderately strong stimulus gave rise to renewed pulsation which persisted for a certain length of time, even on the cessation of stimulus.

THE LATENT PERIOD

Having thus found that a leaflet in a state of standstill can be made to give response to instantaneous stimulation, it is possible to determine the latent period. We have seen that the latent period of the rhythmic cardiac tissue is,

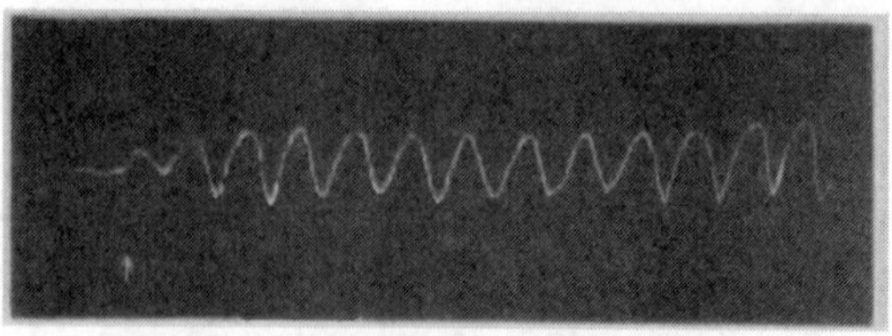

FIG. 152.—Multiple responses, under moderate tetanisation, in leaflet originally at standstill.

generally speaking, longer than that of the ordinary skeletal tissue. In the plant, again, while the average value of the latent period of ordinarily responding *Mimosa* is ·1 second that of the multiple responding *Biophytum* is ·4 second.

The latent period of motile leaflet of *Desmodium* is of the same order. It varies from ·4 second to ·5 second.

THE APEX TIME

In order to study in detail the time-relations, I took on a fast-moving plate the response of *Desmodium* leaflet, originally in a state of standstill, to the action of induction-shock. The frequency of oscillation of the recording-plate was once in a second, hence successive dots represent intervals of

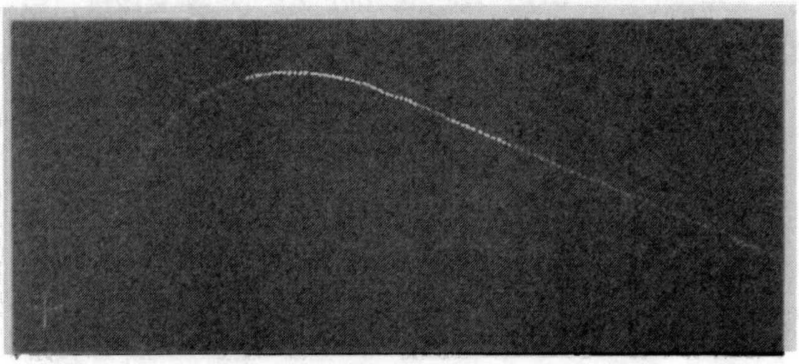

FIG. 153.—Response of *Desmodium* leaflet originally at standstill. Successive dots at intervals of 1 second.

1 second. The entire response, consisting of contractile down-movement and subsequent recovery, was accomplished in the course of 2 minutes and 45 seconds; the leaflet attained its maximum down or contractile position 45 seconds after the application of stimulus. The top of the response-curve is seen to be somewhat flattened, indicating a persistent contraction from which recovery takes place slowly. The period of relaxation is much longer, being 120 seconds; the record in fig. 153 shows recovery which is not quite complete.

Taking thus a typical case we find that in *Desmodium*,

(1) The latent period is ·5 second.

(2) The apex time is 45 seconds.

(3) The period of relaxation is 120 seconds.

The Refractory Period

A striking characteristic of the rhythmic cardiac tissue of the animal is its long refractory period. It is found that by applying successive stimuli there is a minimum resting-interval, the diminution of which brings about such a loss of excitability as to abolish response. This minimum interval is known as the Refractory Period, because short of this period the tissue takes no account of stimulus or is refractory to it. In the cardiac muscle the refractory period lasts during the entire period of the contraction.

In the response of the leaflet of *Desmodium* also we observe in this respect a very striking similarity, in its

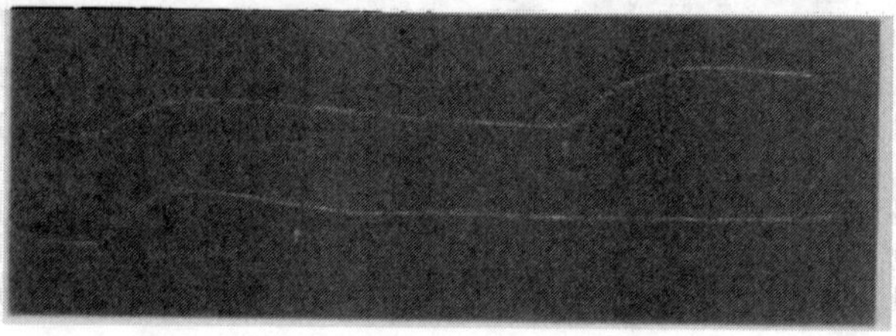

Fig. 154.—Record demonstrating refractory period in *Desmodium*. In the lower record the second stimulus, applied after 45 seconds, is seen to be ineffective, having fallen within the refractory period. In the upper record the second stimulus, applied after an interval of 90 seconds, is seen to be effective.

possession of a long refractory period. This will be seen in fig. 154, where the leaflet in a state of standstill was subjected to two successive stimuli of induction-shocks. In the lower record of the two, after the responsive movement due to the first stimulus, a second was applied after an interval of 45 seconds. The leaflet, however, took no account of it, there being induced no second response. The refractory period in this case is seen to be longer than 45 seconds. After a suitable interval the same leaflet was subjected to two successive stimuli, and at an interval of 90 seconds. It is seen from the upper record that the second stimulus was effective, having fallen on the leaflet beyond the refractory limit.

I stated in a previous chapter that, strictly speaking, there is no such thing as spontaneous movement; that the energy which expresses itself in the pulsating activity of the plant is in reality derived from external sources. It is thus the stimulus supplied by the environment which is held latent by the plant-tissue to find expression later in rhythmic pulsations. For the experimental demonstration of this theory we may carry out the following investigations:

(1) The effect of stimulus on the pulsating activity of a specimen in a moderately sub-tonic condition.

(2) The effect of stimulus on a leaflet in a state of standstill. The renewal of pulsation by absorption of energy has already been shown in the records given in figs. 150, 151, 152, and 153.

(3) Observation of the effect of gradual depletion of energy on the pulsation of the leaflets.

(4) Effect of fresh stimulation on specimens in which the stored energy has been allowed to run down.

Effect of Stimulus on Sub-tonic Specimens

We take a specimen in which the pulsating activity is moderate or feeble, and subject it to stimulation. If the rhythmic activity is the result or after-effect of stimulus previously absorbed, then a more vigorous pulsation may be expected from the greater accession of energy.

This inference is found fully verified in the record given in fig. 155. The specimen had been kept in the dark, and its pulsating activity was only moderate as seen in the first part of the record. The leaflet was then subjected to the stimulus of light from a Nernst electric lamp for half an hour. The record taken after this interval shows a marked enhancement in the amplitude of pulsation.

Effect of Gradual Depletion of Energy

Having observed the effect of accession of energy, it is still more interesting to note the effect of the converse process of gradual depletion of energy. I had hitherto been relying on chance specimens for exhibition of the condition of standstill. As the previous history of the particular specimen was not known from moment to moment, no definite information was available as to the cause of the stoppage of pulsation, except the very natural inference that it must have been due to the run-down of the stored energy. In order to verify this inference, I now proceeded deliberately to isolate an active specimen from all accession of energy from outside, and observe the effect of gradual depletion of energy that had been stored up. The cut specimen, mounted in the usual

FIG. 155.—After-effect of stimulation on pulsation of *Desmodium* leaflet in sub-tonic condition.

manner, was kept in a dark room, and a continuous record of its pulsations was taken all the time.

In taking a series of such records under isolation, it was found that the persistence of the rhythmic activity depended on the vigour of the specimen—that is to say, on the storage of energy in the tissue. Thus a vigorous specimen exhibited persistent activity for more than twenty hours ; its pulsations showed great uniformity for the first twelve hours, after which the amplitude began to decline and the rhythmic beat came to a stop at the twenty-first hour.

In another specimen, less vigorous, the pulsations of the isolated specimen came to an end at the ninth hour. A continuous record from the third to the sixth hour is seen in fig. 142, given on page 294; the series of records are to be read from below to above each series, lasting for one hour and twenty minutes. It is seen that the amplitude is

FIG. 156.—Effect of depletion of energy on the pulsation of isolated specimen of *Desmodium gyrans.*

practically constant; the period is also constant in the first two series, there being twenty-seven pulsations in each in the course of eighty minutes. The period is, however,

FIG. 157.—Gradual stoppage of pulsation in isolated leaflet of *Desmodium*.

slightly lengthened in the third series, there being twenty-six pulsations in the given time. The record was continued (fig. 156) during the seventh, eighth, and part of the ninth hour. In these two series, the effect of the depletion of stored energy is well seen in the regular diminution of the amplitude and the prolongation of the period; there are now, in the lower or earlier series, twenty-three pulsations in the course of eighty minutes, instead of twenty-seven, which was found at the beginning. The period becomes still more prolonged in the next series, where there are only twenty pulsations. The pulsating activity of this specimen was found completely arrested at the end of the ninth hour.

In another specimen less vigorous, the arrest took place at the seventh hour. I give a record taken during the seventh hour (fig. 157) which shows the process of arrest in a very interesting manner.

Renewal of Pulsation under Fresh Stimulus

The arrest of pulsation seen here is not due to death or loss of sensibility, but merely to the run down of stored energy. I shall presently show that the activity of the leaflets can in these cases be renewed by the incidence of an external stimulus. Before describing the experiments, it is well to anticipate the modifying results brought on by the varying loss of stored energy. The extent of depletion will, it is obvious, depend on the length of time during which the specimen had been kept isolated from the external supply. At the moment of arrest of pulsation the specimen will still have a certain reserve, but not enough to cause an overflow. At this stage a stimulus of even short duration is likely to give rise to multiple responses. If we allow a longer time to elapse after the stoppage of pulsation, then the storage-level will be much lowered. A more intense stimulus or one of longer duration will now be effective; the response is more likely to be single rather than multiple. And lastly, if we allow a very long time to elapse after the cessation of pulsation, the vitality will be found to have so far declined as to bring about the condition of death. To recapitulate: the effect of fresh incidence of stimulus on a leaflet brought to a state of standstill by isolation will be modified according to the period which had been allowed to elapse after the cessation of pulsation. Stimulus applied within a short time of stoppage will give rise to multiple responses ; after a long period, an identical stimulus will be found far less effective. After a still longer interval, which is critical, even a strong stimulus will fail to evoke any response.

These theoretical considerations will be seen verified in the following series of records. In fig. 158, stimulus of light of two seconds' duration was applied after half an hour of the stoppage of the pulsation of the leaflet. The response is

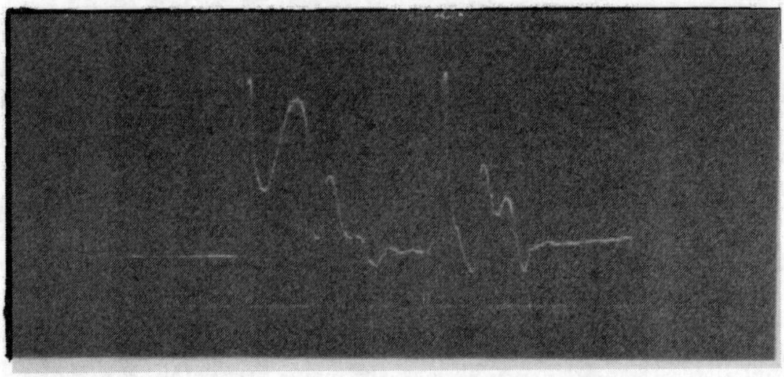

FIG. 158.—Effect of stimulus in renewing pulsation of *Desmodium* brought to standstill through depletion of energy. Note two series of multiple responses due to stimulus of light of two seconds' duration.

seen to be multiple even under a stimulus of this short duration. The stimulus was applied once more, with similar results.

In the next series of records, stimulus was applied on a specimen which had been in a state of standstill for five hours. The store of energy here had undergone a considerable decline. Stimulus of light of two seconds' duration was found ineffective ; it was only after an application lasting for half a minute that a response took place (fig. 159). The response was single, instead of multiple as in the last case. Stimulus was applied a second time, but of still longer duration of one minute. The response is now found to consist of a large, followed by two small, pulsations.

I next took another series of records, with a specimen which had been in a state of standstill for seven hours.

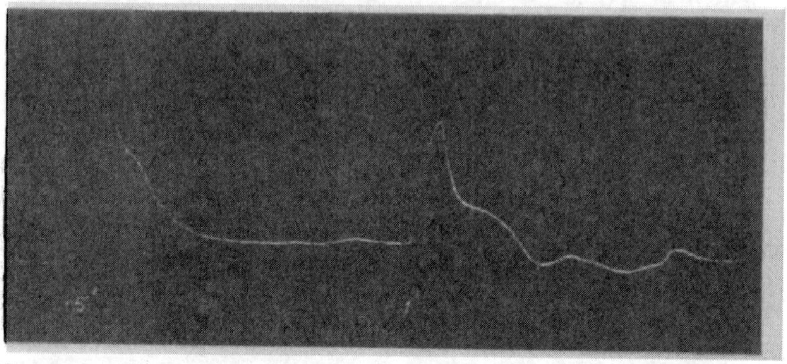

FIG. 159.—Response of *Desmodium* leaflet at standstill to stimulus of light of ·5 and 1 minute's duration.

Stimulus of light was successively applied for one, two, three and four minutes. The general result is somewhat similar to those that have been seen in the previous record. The noticeable differences are, first the diminished amplitude

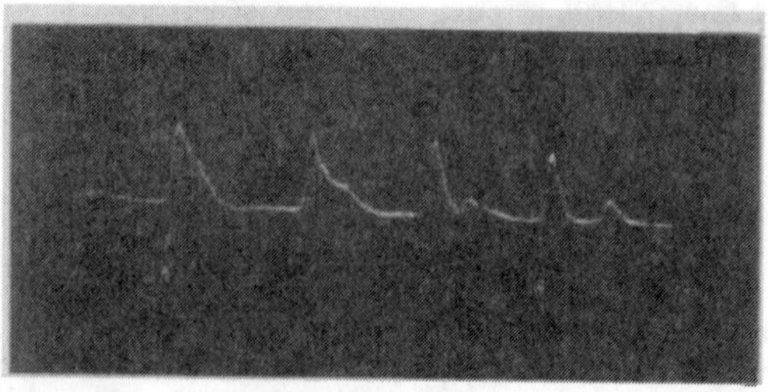

FIG. 160.—Response of a depressed specimen of *Desmodium* to stimulus of light applied successively for one, two, three, and four minutes.

of response, and the exhibition of decline in the successive responses. The specimen was nearing the critical condition of death, and after a while there was a cessation of all response. In other specimens the critical period was found to be exceeded when the state of standstill had been allowed to persist for nine or ten hours. In these, stimulus failed completely to evoke any response.

Summary

The leaflet of *Desmodium* comes to a state of standstill by depletion of its store of energy.

The pulsation may be revived by the action of various stimuli, such as that of light or of induction-shock.

Desmodium leaflet in a state of standstill gives a single response to a single stimulus of induction-shock of moderate intensity. In a typical case the latent period is ·4 second, the apex time 45 seconds, and the period of relaxation 120 seconds. The response-curve exhibits a flattened top.

The response of *Desmodium*, like that of the cardiac tissue, is characterised by a long refractory period.

The rhythmic activity of the leaflet of *Desmodium* comes to an end when its store of energy is depleted. A leaflet isolated from external sources of stimulation is thus gradually brought to a state of standstill.

In this condition, response occurs under fresh stimulation. If the depletion of energy had not been excessive, then a moderate stimulus gives rise to multiple responses. But under greater depletion even a very strong stimulus induces only a single response.

CHAPTER XXIV

EFFECT OF ELECTRIC STIMULATION ON THE PULSATION OF *DESMODIUM GYRANS*

Effect of electric shock on *Desmodium* leaflet—Incapability of tetanus —Extra pulsation induced by electric shock—Relative effectiveness of electric stimulus at diastolic phase—Effect of transmitted excitation on normal pulsation of heart and on pulsation of *Desmodium* leaflet—Effects of acceleration and inhibition.

IN the previous chapter we studied the action of electric shock on the leaflet of *Desmodium* in a state of standstill. We will now inquire into the effect of induction-shock on the pulsating leaflet.

The cardiac tissue is known to be incapable of tetanus.

FIG. 161.—Application of strong tetanising shock at arrow brings about diminished amplitude of pulsation.

This is also true of the rhythmic tissue of *Desmodium.* Under tetanising shocks there is no continuous contraction. Under feeble stimulus the amplitude of pulsation may in certain circumstances be enhanced. Under excessive tetanisation the natural pulsations, generally speaking, become irregular or diminished in amplitude (fig. 161).

Extra Pulsation due to Induction-Shock

In the case of cardiac tissue it is known that the excitability is least at the commencement of systole. A moderate stimulus applied at this period induces hardly any effect. During the occurrence of diastole, however, excitability is relatively great. Hence, if stimulus be applied at diastole, it is followed by an extra contraction.

Turning to *Desmodium*, I find that similar characteristics obtain in this case also. The application of stimulus at the beginning of the contractile movement, which corresponds to systole, has little or no effect. But during the movement of

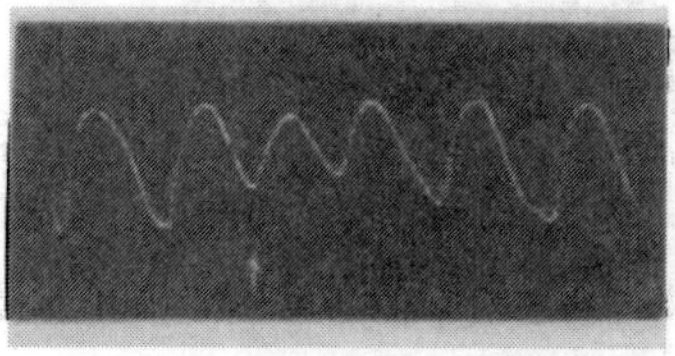

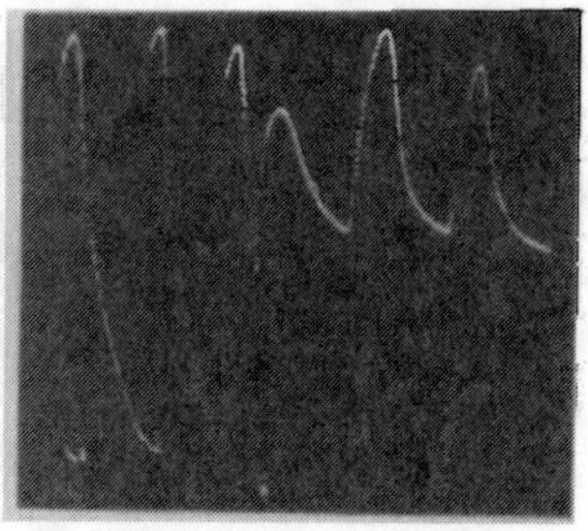

Figs. 162, 163.—Extra pulsation induced by induction-shock applied at diastolic phase.

relaxation, which corresponds to diastole, the application of an induction-shock induces the arrest of relaxation and gives rise to an extra or interpolated pulsation. I give here two records, out of many, obtained from two different specimens, in both of which a momentary electrical shock was applied about the middle of the phase of diastolic relaxation. The arrest of relaxation, and reversal to the opposite phase of contraction, giving rise to an extra pulse, are here clearly seen in both cases (figs. 162, 163).

Effect of Transmitted Excitation on Pulsation of *Desmodium*

On exciting the nerve that goes to the heart, two opposite effects have been observed. In some cases there is induced a diminished amplitude of pulsation: in other cases an augmentation. It has been supposed that this nerve contains two different kinds of fibres, the accelerators and the inhibitors. In the case of lower animals, such as the tortoise, these have been isolated by Gaskell. But the accelerator fibres have not been distinguished in the mammalian vagus. It has been found that, generally speaking, in a vigorously beating heart the effect of vagus-stimulation is to induce depression, whereas in a sluggish heart the same stimulation is apt to induce an augmentation.

To return to the case of the pulsating pulvinule of

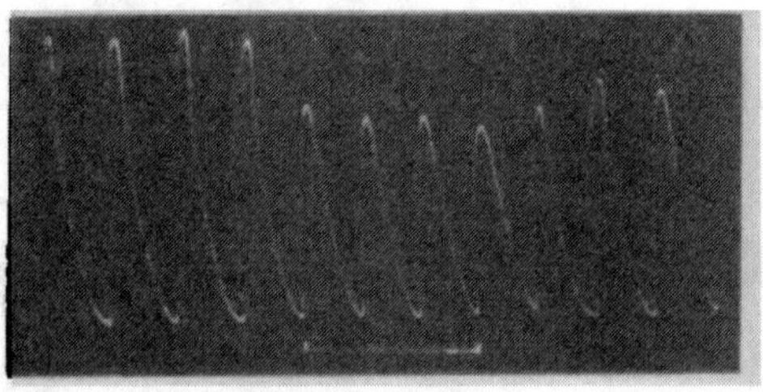

FIG. 164.—Inhibitory effect of transmitted excitation on the pulsation of vigorous leaflet of *Desmodium*. The line below indicates duration of transmitted excitation; note the gradual removal of inhibitory effect on cessation of stimulation.

Desmodium, the question arises whether the organ is in communication with any conducting channel by which distant excitation might be transmitted to it. The fact of such transmission, if it occurred, would be tested by its modifying influence upon the normal pulsation. In order, then, to subject this question to the test of experiment, I made suitable electrical connections—one contact being on the petiolule, 5 mm. below the contractile pulvinule, and the other still lower down on the petiole. Normal responses were first taken, after which indirect stimulation was applied by means of tetanising electrical shocks of moderate intensity. The first specimen employed was very vigorous, as will be observed from the amplitude of its pulsations (fig. 164). It will be noticed that stimulation, thus applied at a distance, was transmitted and induced an inhibitory effect in diminishing the amplitude of normal pulsation. On the cessation of excitation the pulsations are seen gradually to regain their normal amplitude. This inhibitory effect is what takes place more frequently, and may be regarded as typical. Somewhat exceptional is the converse effect of augmentation, seen in the next record (fig. 165). The particular specimen was less vigorous; this fact is seen in the smaller amplitude of its normal pulsations, the magnification being the same in the two cases. After these normal pulses had been

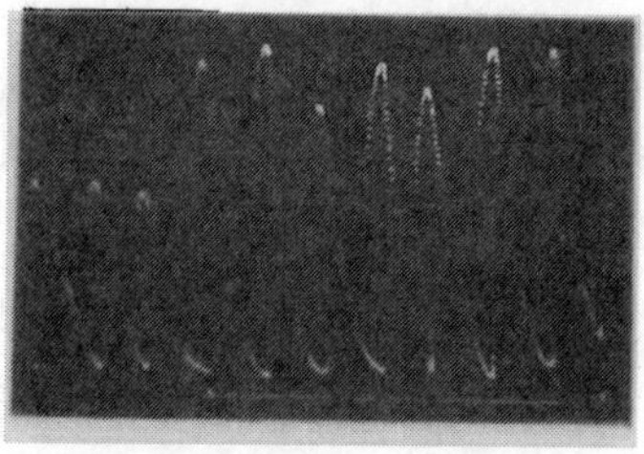

FIG. 165.—Augmented pulsation induced by transmitted excitation in less vigorous specimen of *Desmodium*.

recorded, indirect stimulation of moderate intensity was applied, as in the previous case. It will be noticed that, in consequence of this, there occurred a marked enhancement of the amplitude of pulsation. Even on considerably raising the intensity of the stimulation this enhancement of pulsation still persisted.

A remarkable parallelism has thus been shown to exist between the responsive characteristics of rhythmic animal and vegetal tissues. This is seen in their incapability of tetanus, in their prolonged refractory period, in the extra pulsation induced by electric shock at the diastolic phase, and in the transmitted excitation causing in different circumstances effects either of inhibition or acceleration. Other similarities, equally remarkable, will be found in the effect of temperature and of drugs on the pulsating activity of animal and vegetal tissues.

Summary

The rhythmic tissue of *Desmodium*, like the rhythmic cardiac tissue, is incapable of tetanus.

Pulsating leaflet of *Desmodium*, like the pulsating heart, is more susceptible to excitation at diastole than at systole. An extra pulsation is induced by an electric shock applied during the diastolic phase.

Transmitted excitation affects the normal pulsations of rhythmic tissues—animal or vegetal—in a similar manner. In certain circumstances the effect is one of inhibition; in other circumstances the effect is one of acceleration.

CHAPTER XXV

EFFECT OF TEMPERATURE ON RHYTHMIC PULSATION OF *DESMODIUM GYRANS*

Effect of lowering of temperature on rhythmic pulsation of cardiac tissue—Similar effect on the pulsation of *Desmodium*—Increase of systolic limit during cooling—Minimum temperature for arrest of pulsation—Arrest by cooling and subsequent revival by warming—Increase of diastolic limit during warming—Effect of rise of temperature on the pulsation of frog's heart—Similar effect on the pulsation of *Desmodium*—Effect of rise above and return to normal temperature—Diminution of systolic contraction during rise of temperature—Increase of systolic contraction during fall of temperature—Permanent arrest due to heat-rigor.

In the course of our study of automatic pulsation of *Desmodium* we shall find striking similarities between the rhythmic activities of the plant and animal tissues. In the present chapter we will study in detail the effect of temperature in modifying the amplitude and period of the spontaneous movements.

I have already described the thermal chamber by means of which the temperature of the plant can be regulated. Temperature, as we have seen, may be raised to any degree by the adjusting of the heating current which passes through a coil of German silver. Lowering of temperature, on the other hand, is effected by allowing a stream of cooled air to pass through the chamber containing the specimen.

Effect of Lowering of Temperature

In the rhythmic pulsation of frog's heart the marked effect of variation of temperature is to change the period and modify the amplitude of pulsation. Lowering of temperature has the effect of lengthening the period and enhancing the amplitude. This is seen in the following record (fig. 166), where the normal pulsations are modified in consequence of lowering the temperature through a few degrees.

The following experiment, carried out on the leaflet of *Desmodium*, shows that the effect of lowering of temperature on it is precisely the same. The temperature of the room at the time of the experiment was 30° C. A record of three normal pulsations was taken at this normal temperature.

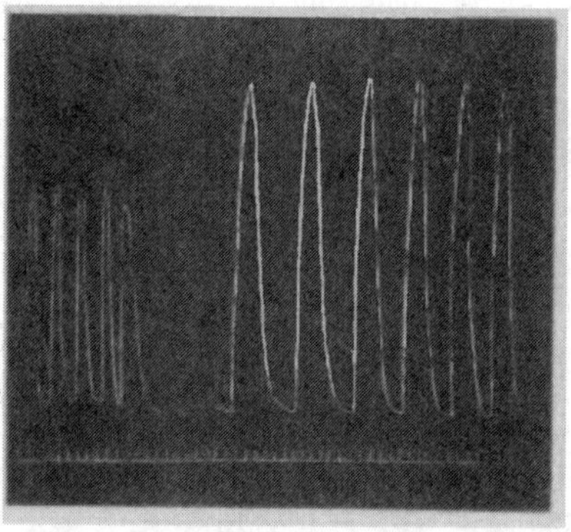

FIG. 166.—Effect of lowering of temperature in increasing amplitude, and decrease of frequency of pulsation of frog's heart. Series to the left represent normal pulsations at the temperature of room; series to the right were recorded at a temperature several degrees lower. (Brodie.)

The specimen was then gradually cooled by sending through the chamber a stream of cold air, the record being taken all the time. The successive dots in the diagram are at intervals of 2 seconds. Hence the period of complete pulsation can be accurately determined.

It will be seen from fig. 167 that at 30° C. the period of a complete pulsation was 80 seconds, the amplitude being 25 mm. On lowering of temperature to 29° C. the period became lengthened to 88 seconds, the amplitude being enhanced to 30 mm. At 28° C. the period was protracted to 96 seconds, the amplitude being enhanced to 35 mm. And finally at 27° C. there was further lengthening of the period to 110 seconds, the amplitude of pulsation being enhanced to 40 mm.

In addition to the changes of period and amplitude there is another noticeable effect induced by variation of temperature. We have seen in the case of *Mimosa* that the lowering of temperature induces a depression of the leaf, and the rise of temperature an erection of the leaf. Effects similar to

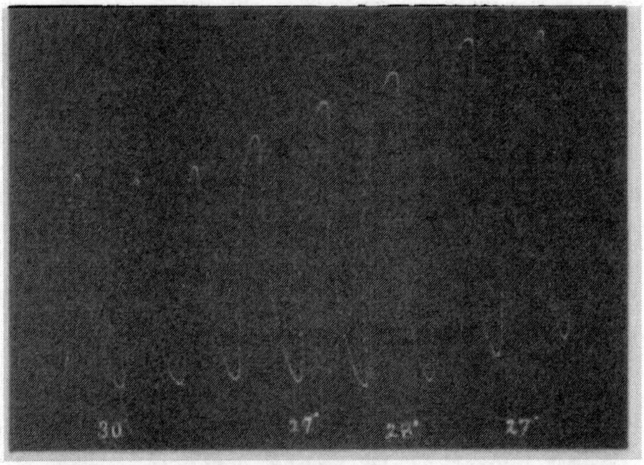

FIG. 167.—Effect of lowering of temperature on the amplitude and frequency of *Desmodium* pulsation. Note the shifting of base-line upwards, indicating fall of leaflet during cooling; observe also the enhancement of systolic contraction.

these are also induced in the leaflet of *Desmodium*. The result of cooling is thus a slight fall of the leaflet, in consequence of which we observe the shifting of the base-line upwards. A still more striking effect is the increase in the extent of contraction, by which the systolic limit is enhanced. This will clearly be noticed in the series of records in fig. 167.

The rhythmic pulsation of the cardiac tissue is arrested when subjected to a certain low temperature. Similarly the pulsation of *Desmodium gyrans* is arrested at a sufficiently low temperature. The critical point is somewhat modified by the tonic condition of the specimen. With vigorous specimens the temperature at which arrest takes place may be as low as 17° C.

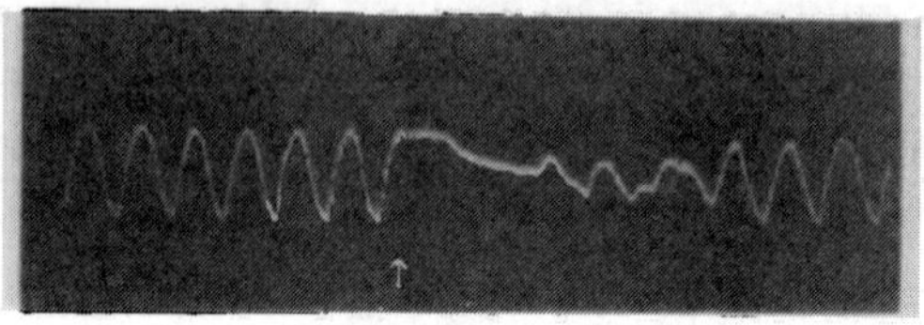

FIG. 168.—Effect of rapid cooling by ice water applied at the moment marked by arrow. Note arrest at systole and gradual revival on warming, the pulsations exhibiting staircase increase.

There are certain interesting points in connection with the arrest brought about by cooling. Two series of records are here given in which the arrest was quickly brought about

FIG. 169.—Magnified record given by a different specimen of arrest of pulsation by cooling and subsequent revival by warming. Note extension of diastole during warming.

by the application of cold water to the pulvinule. The leaflet was then allowed to return to the temperature of the room, with the revival of pulsatory activity. The entire process of arrest and subsequent revival is clearly seen in fig. 168. It will be noticed that sudden cooling arrested the pulsation at systole and that on gradual warming the rhythmic activity was revived with the gradual restoration of the original amplitude.

This is still better seen in fig. 169, which gives a magnified record of the arrest due to cold, and subsequent restoration of pulsation on return to the temperature of the room. Ice-cold water was applied after the second pulsation. The arrest at systolic contraction is clearly

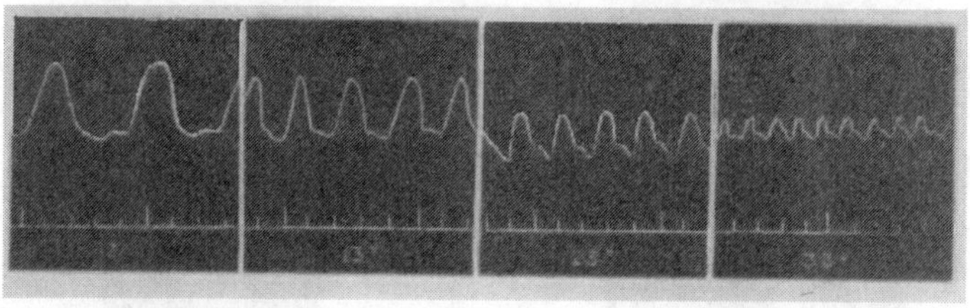

FIG. 170.—Effect of rise of temperature on the amplitude and frequency of pulsations of the heart of frog. (Pembrey and Phillips.)

demonstrated. As the temperature was raised there was increasing expansion, in consequence of which the diastolic excursion was continuously increased. The effect of cooling is thus to increase the force of contraction and diminish that of expansion. The effect of warming is the reverse of these.

EFFECT OF RISE OF TEMPERATURE

We have seen that both in the rhythmic animal and vegetable tissues, the period is increased and amplitude enhanced under the lowering of temperature. The converse is the case with rise of temperature. Fig. 170 shows the effect of rising temperature on the pulsation of the heart of the frog. Similar effects are induced in the pulsation of *Desmodium*. Fig. 171 gives a series of pulsations of the leaflet at the temperatures of 19° C., 23·5° C., and 28·5° C., the record in each case being continued for a period of 20 minutes. At a low temperature the pulsation is apt to be somewhat irregular, hence the unequal amplitude in the successive pulsations at 19° C.

It is apparent that while at 19° C. there were $3\frac{1}{2}$ pulsations, at 23·5° C. the number had been increased to $4\frac{1}{2}$, and at 28·5° C. to 6 pulsations.

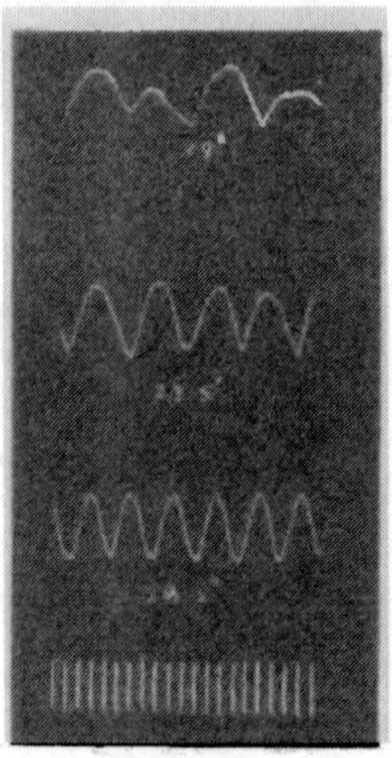

FIG. 171.—Effect of rise of temperature on the pulsation of *Desmodium gyrans*. Time-marks below indicate intervals of 1 minute.

FIG. 172.—Effect of rise of temperature on a different leaflet.

Taking a more vigorous specimen, I obtained records for 12 minutes each at temperatures of 28·5° C., 31·5° C., and 34·5° C. It will be seen that while at 28·5° there were only 4 pulsations, these had become increased to $6\frac{1}{2}$ pulsations at 31·5° and to 10 pulsations at 34·5° C. The other noticeable feature is the marked diminution of amplitude with the rise of temperature (fig. 172).

Instead of taking isolated records at different temperatures, I next raised the temperature of the plant-chamber very gradually, by careful manipulation of the heating current, and obtained a record with a different specimen. In this way the temperature was raised continuously from

30° C. to 38·5° C. How regularly the frequency of pulsation is increased, and the amplitude diminished, with the rise of temperature is shown in fig. 173.

In the record just given there is observable an arrest of pulsation when the specimen was subjected to as high a temperature as 38° C. But by accustoming it to warmth, the plant can resist even higher temperatures. Thus I kept

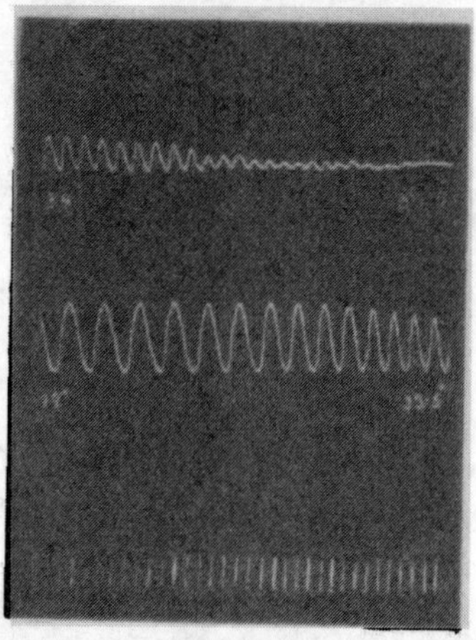

FIG. 173.—Effect of continuous rise of temperature from 30° C. to 38·5° C.

several specimens in a glass-house, the temperature in which at midday was 37° C. These specimens could be exposed to a temperature as high as 45° C. without the arrest of the pulsation. I reproduce here a record (fig. 174) where the specimen was gradually raised from 30° C. to 42° C., and then allowed to cool and return to the temperature of 30° C. It will be seen that the amplitude of pulsation was continuously decreased, yet there was no arrest even at 42° C. On cooling, the amplitude was restored to the original value.

I have shown that under excessive cooling the force of expansion is reduced, in consequence of which there is an arrest towards systole. With excess of heat, on the other hand, the reverse effect takes place. On account of increase of force of expansion, or diminution of force of contraction, the systolic limit is progressively diminished. When the

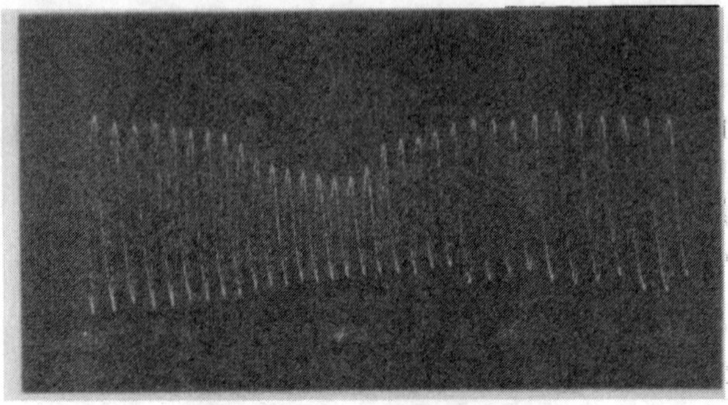

FIG. 174.—Effect of continuous rise from 30° C. to 42° C. and return to 30° C. Note progressive diminution of systolic contraction during warming, and its increase during cooling.

specimen is allowed to cool, the reverse effect is exhibited by gradual enhancement of systolic contraction. Similar

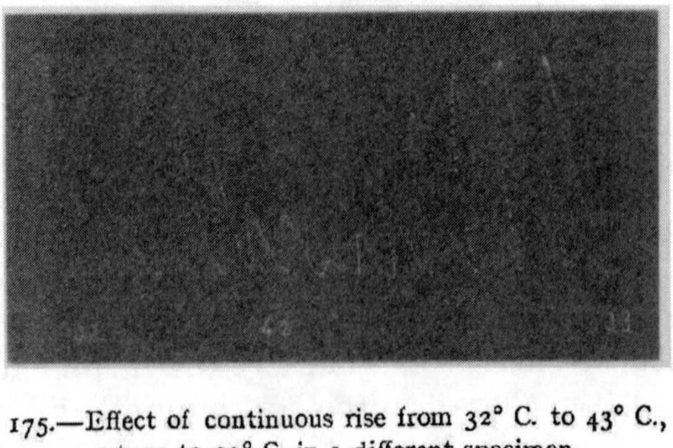

FIG. 175.—Effect of continuous rise from 32° C. to 43° C., and return to 33° C. in a different specimen.

effects are seen to take place in a record (fig. 175) obtained with a different specimen.

The temperature may sometimes be raised to 45° C. without inducing any arrest of pulsation. A tendency is now observed towards contraction, as shown by the general shifting of the pulsations upwards. This movement of contraction goes on till there is a complete arrest brought about by heat rigor.

SUMMARY

The effect of lowering of temperature on the rhythmic pulsation of *Desmodium gyrans* is similar to that on the pulsation of frog's heart. Lowering of temperature enhances the amplitude but reduces the frequency of pulsation of both.

The pulsation of *Desmodium* leaflet is arrested at the minimum temperature of about 17° C. Arrest takes place at systole; gradual warming revives the pulsation, which undergoes a staircase increase with enhancing diastolic expansion.

Rise of temperature induces enhanced frequency and diminished amplitude of pulsation.

During rise of temperature to about 43° C. there is a tendency of arrest towards diastole. The systolic contraction undergoes continuous diminution during rise of temperature.

During the fall of temperature there is a gradual enhancement of systolic contraction.

The temperature maximum at which arrest of pulsation takes place may be as high as 45° C. Above this temperature there is a tendency to contraction and permanent arrest under heat-rigor.

CHAPTER XXVI

EFFECT OF CHEMICAL AGENTS ON THE AUTOMATIC PULSATION OF *DESMODIUM GYRANS*

Application of gaseous or liquid reagents—Modifying influence of, tonic condition of specimen, strength, and duration of application—Effect of sugar solution—Effect of alcohol—Action of carbonic-acid gas—Effects of anæsthetics, ether, and chloroform—Action of carbon disulphide—Effect of copper sulphate solution—Effect of potassium cyanide solution—Antagonistic actions of acids and alkalis on the pulsations of the heart and of *Desmodium*—Similarities of reaction in rhythmic tissues, animal and vegetal.

WE will now study the effects of various chemical agents on the pulsating activity of the *Desmodium* leaflet. These may be applied either externally or internally. As regards external application, a liquid reagent may be applied directly on the pulvinule; gases and vapours, on the other hand, are made to circulate in the plant-chamber. We may secure internal application by forcing in the solution at the cut end of the petiole, by means of hydrostatic pressure (cf. fig. 141).

The characteristic effects of different reagents are seen exhibited in the induced change of period or of amplitude of pulsation. Another noticeable effect often observed is the transposition of the systolic or diastolic limit of pulsation. Certain agents may thus induce an arrest at systole, others at diastole.

The effect of a given agent may be modified by various conditions, such as internal or external application, the strength of the solution, and the duration of application.

Thus the same agent which in a strong solution induces depression, may in a very dilute solution cause an exaltation.

The result also is dependent on the tonic condition of the tissue. A specimen in which the amplitude of pulsation is the maximum possible cannot exhibit any further increase under a stimulating agent. But with a less vigorous specimen the effect of the same agent is manifested by a marked

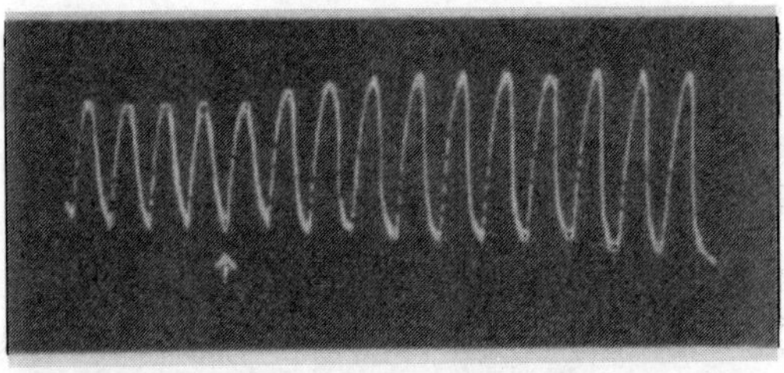

FIG. 176.—Stimulating action of dilute sugar solution applied at the moment marked by an arrow.

enhancement of amplitude of pulsation. Again, a vigorous plant may survive a given dose of a toxic agent; while a less vigorous specimen will succumb under the same treatment. Finally we have the interesting phenomenon in

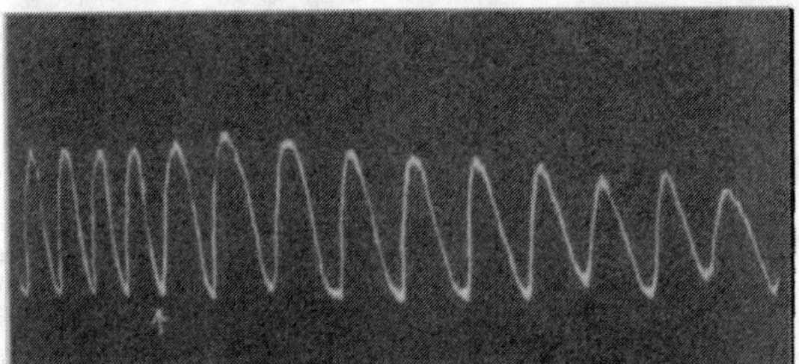

FIG. 177.—Effect of external application of alcohol. Preliminary enhancement followed by depression, the period of pulsation being prolonged.

virtue of which the plant accommodates itself to a changed unfavourable condition.

Certain chemical agents act as stimulants, inducing an enhancement of amplitude of pulsation. As an example of this I may mention dilute sugar solution. The specimen experimented on was but moderately vigorous. After taking four normal pulsations, which are seen to be uniform (fig. 176), a 2 per cent. solution of sugar was applied internally at the moment marked by an arrow. No immediate effect was noticeable, but after an interval of a single

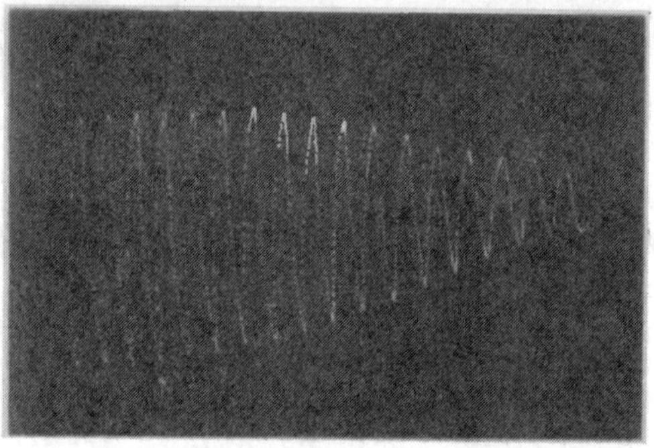

FIG. 178.—Effect of internal application of 15 per cent. solution of alcohol.

FIG. 179.—Effect of dilute carbonic-acid gas; enhancement of amplitude and slowing of the period.

pulsation the stimulating character of the agent became evident by the resulting staircase enhancement of pulsation.

EFFECT OF ALCOHOL

The effect of this reagent in dilute solutions is in general to induce an enhancement of response. Stronger solution, however, induces a depression which may culminate in arrest of pulsation. In fig. 177 a slight preliminary exaltation of amplitude, followed by depression, may be noticed. The period of succeeding pulsations is found to become prolonged.

The effect of internal application of alcohol is very similar to that of external application. In fig. 178 is seen a

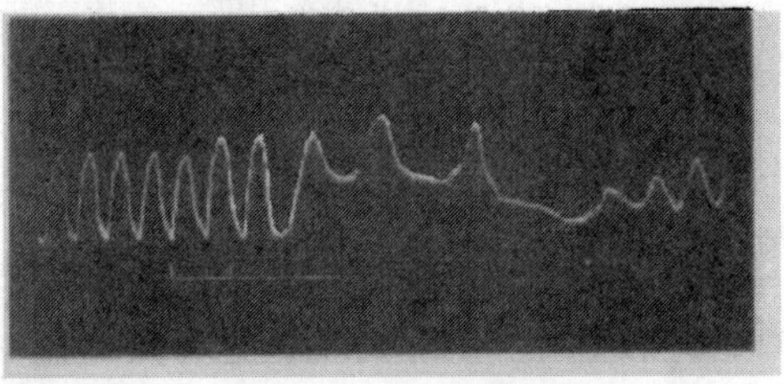

FIG. 180.—Strong carbonic-acid gas, inducing arrest. Line below indicates duration of application. Slow revival of pulsation on substitution of fresh air.

FIG. 181.—Effect of internal application of water charged with carbonic acid.

record exhibiting the action of a 15 per cent solution of alcohol.

EFFECT OF CARBONIC-ACID GAS

This gas when diluted with air causes an enhancement of amplitude, though the period becomes longer (fig. 179).

Application of undiluted gas, however, induces an arrest. If fresh air be now substituted, there is produced a slow revival of pulsation (fig. 180).

In another experiment I obtained a record of the effect of internal application of water charged with carbonic acid. The pulsating activity was found slowed down, the amplitude of the pulsation also undergoing a diminution. The depressing effect is, however, seen to pass away gradually (fig. 181).

EFFECT OF ETHER

This agent was applied in the form of vapour, which was slowly blown into the plant-chamber. If the vapour be

FIG. 182.—Effect of vapour of ether; depression and subsequent arrest. Pulsations revived on blowing off the vapour.

much diluted with air, then the first effect of ether is to induce a transient exaltation, followed by depression and arrest of pulsation. If the leaflet be subjected to strong vapour, or if the application be prolonged, then the arrest of pulsation proves to be permanent. But if diluted vapour is employed and fresh air substituted immediately after the arrest, then there is a slow revival of pulsation. This can be seen in fig. 182, where after the application of ether an arrest took place after three rapidly diminishing pulsations. On blowing off the ether vapour, the pulsation is seen to revive slowly after a period of 20 minutes.

EFFECT OF CHLOROFORM

The effect of this reagent on the pulsation of *Desmodium* is similar to that of ether. It is, however, far more toxic in its reaction, a slight excess in the application being attended by permanent arrest of pulsation. In the experiment of which a record is given (fig. 183), diluted chloroform vapour was introduced into the chamber. This is observed at

FIG. 183.—Effect of vapour of chloroform. Immediate excitatory effect followed by depression and arrest. On blowing off the anæsthetic, pulsations were revived after half an hour, represented by a gap in the record.

first to have an excitatory effect in the first two pulsations. The amplitude of the third pulsation became much reduced,

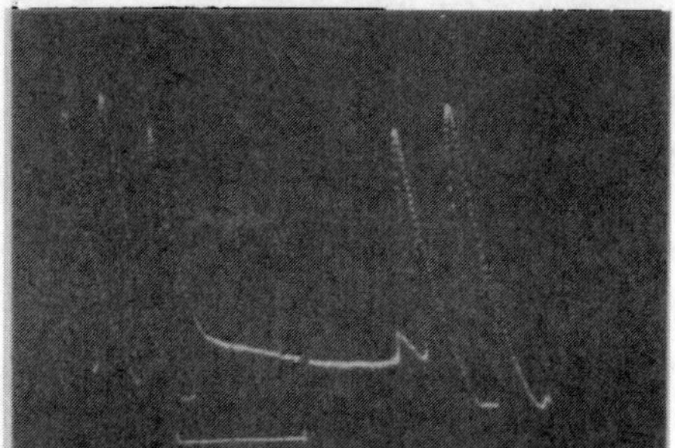

FIG. 184.—Arrest of pulsation by vapour of carbon disulphide, and revival after readmission of fresh air. Duration of application indicated by the horizontal line under the record.

and an arrest ensued at the fourth pulsation after the application of the anæsthetic. The vapour was then quickly blown away and fresh air substituted in the chamber. But the arrest persisted for half an hour. After this interval there was a revival, and the pulsation attained for a time an amplitude even greater than the normal.

Effect of Carbon Disulphide

The vapour of this reagent also arrests the pulsating activity of the leaflet. The record (fig. 184) shows the quick arrest after a single pulsation. Substitution of fresh air is seen again to revive the pulsation.

Effect of Copper Sulphate Solution

As an example of toxic agents we may take solutions of substances like copper sulphate. When a strong solution of

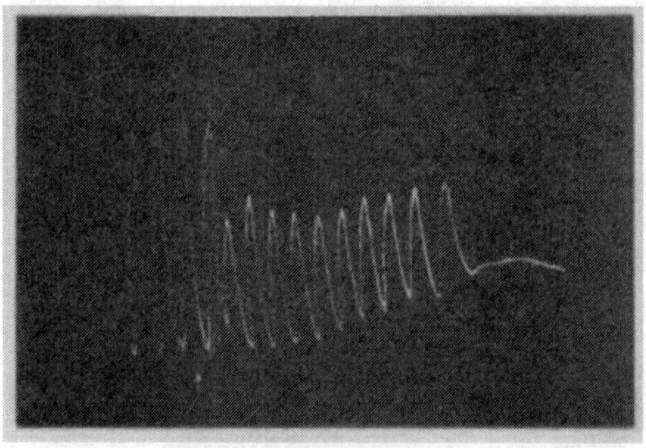

Fig. 185.—Effect of internal application of $CuSO_4$ solution in inducing arrest of pulsation.

this substance is applied directly on the pulvinule, an arrest takes place within a short time. The effect is delayed when the solution is applied at the cut end of the petiole. A record of the effect of internal application of copper sulphate solution is given in fig. 185.

Effect of Potassium Cyanide Solution

We saw in Chapter XII. that the toxic effect of potassium cyanide in abolishing the conductivity of the tissue was far more pronounced than that induced by copper sulphate. The poisonous action of cyanide is equally powerful in abolishing the rhythmic activity of the *Desmodium* leaflet.

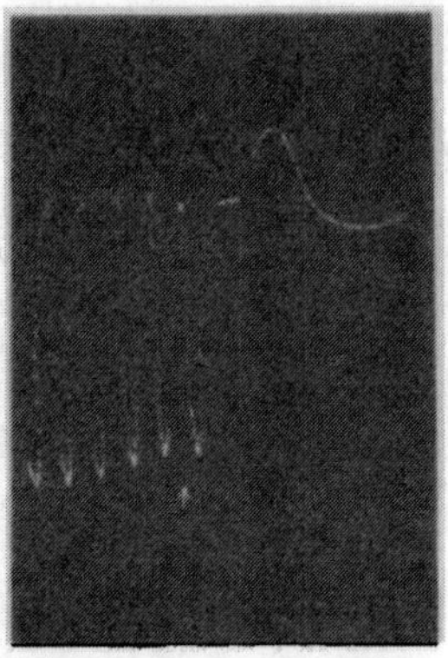

FIG. 186.—Quick arrest of pulsation at systole by the action of KCN solution.

This is well observed in fig. 186, where the pulsation is seen to be quickly arrested at systole.

ANTAGONISTIC ACTIONS OF ACID AND ALKALI

We have hitherto seen the remarkable similarities of the effect of various chemical agents on the rhythmic activities of the animal and vegetal tissues. A very striking

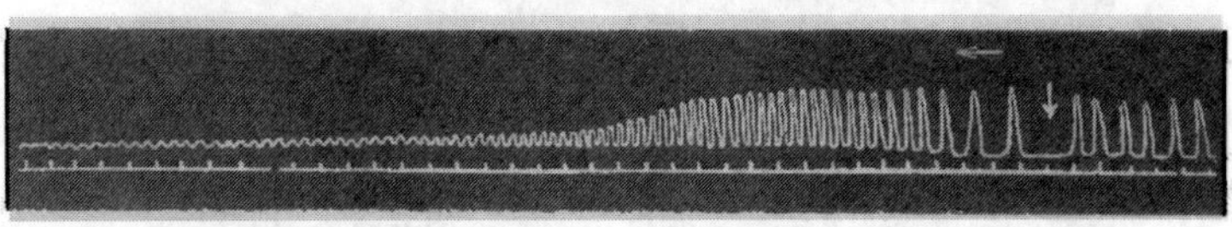

FIG. 187.—Arrest of pulsation of the heart of frog in diastole by the action of dilute lactic acid. (Gaskell.) Record to be read from right to left in this and following figures.

characteristic is the antagonistic reactions of acid and alkali on the animal heart. Application of very dilute acid induces in the heart an atonic reaction, in consequence of which there is induced an arrest of pulsation in the relaxed or diastolic condition (fig. 187). The action of dilute alkaline solution is the very reverse, inducing tonic contraction and arrest in systole (fig. 189).

I find these effects repeated in an astonishing manner in

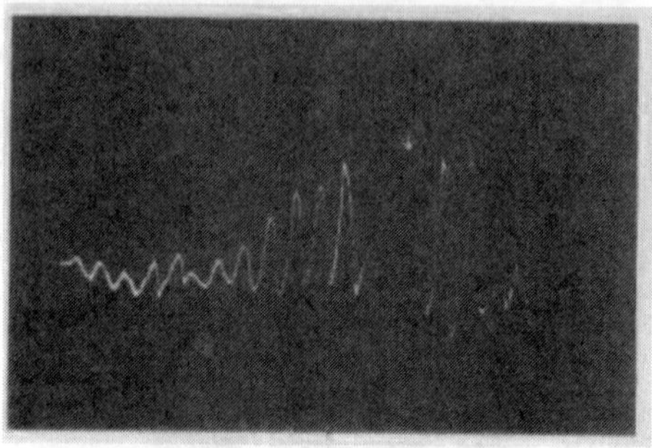

FIG. 188.—Arrest of pulsation of *Desmodium* in diastole, by the action of dilute lactic acid.

FIG. 189.—Arrest of pulsation of heart in systole, by the action of dilute NaHO. (Gaskell.)

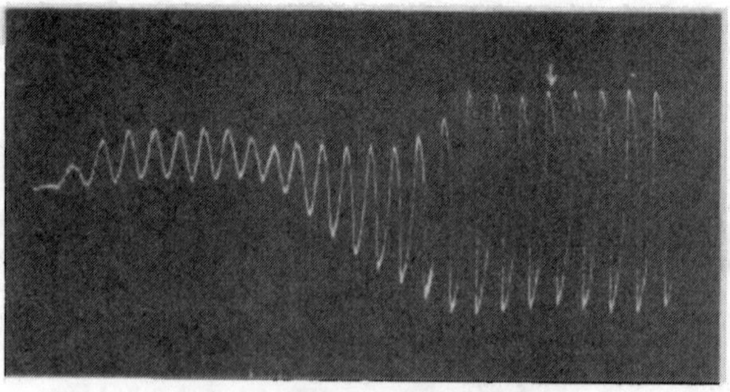

FIG. 190.—Arrest of pulsation of *Desmodium* in systole, by the action of dilute NaHO.

the pulsation of *Desmodium*. The internal application of dilute solution of lactic acid is seen to induce an arrest in a state of diastolic relaxation (fig. 188). The application of dilute NaHO solution (fig. 190), on the other hand, induces exactly the opposite effect—namely, an arrest at systolic contraction.

Summary

The effect of chemical reagent on the pulsating activity of *Desmodium* is to a certain extent modified by the tonic condition of the specimen.

The effect is also dependent on the strength of the reagent. Very dilute application often induces an effect opposite to strong.

Alcohol induces a transient exaltation followed by depression. The period of pulsation is lengthened.

Carbonic acid when dilute induces an exaltation of amplitude, with prolongation of period of each pulsation. Long application of undiluted gas induces an arrest : revival takes place on substituting fresh air.

Vapour of ether induces a temporary arrest of pulsation ; the pulsation may be revived on blowing off the vapour.

The effect of chloroform is similar to that of ether, but its toxic effect is greater, any excess causing permanent arrest. Diluted vapour causes a preliminary exaltation followed by arrest. Revival may take place after quick substitution of fresh air.

Vapour of carbon disulphide applied for a short time also induces a transient arrest.

Copper sulphate solution when applied directly on the pulvinule causes an almost immediate arrest. But when applied at the cut end of petiole, the arrest does not take place till the lapse of a period, required for the solution to ascend to the motile organ.

The poisonous reaction of potassium cyanide solution is more powerful than that of copper sulphate.

The effect of acids and alkalis on the rhythmic movements of *Desmodium* are, as on the animal heart, antagonistic. In both, acids induce a standstill in diastole, while alkalis induce arrest in systole.

CHAPTER XXVII

General Survey

For the study of phenomenon of irritability, the so-called 'sensitive' plants have been selected for the purpose of experimental investigation. From this it must not be inferred that the fundamental reactions that have been demonstrated are different in ordinary plants. By the employment of the electrical mode of investigation, I have shown elsewhere that not sensitive plants alone, but every plant, and also every organ of every plant, is excitable. Even in the matter of mechanical response, longitudinal contraction under excitation may be demonstrated in the case of various organs of ordinary plants; but the extent of movement in such cases is not very great. A 'sensitive plant' differs from others only in the possession of mobile mechanism by which excitatory change of form is manifested by a conspicuous movement.

In the case of *Mimosa*, the responsive down movement or the fall of the leaf is brought about by the differential excitabilities of the upper and lower halves of the pulvinus, the movement being magnified by the long petiolar index. It is these advantageous circumstances that render the motile apparatus of *Mimosa* a good indicator of excitation. The leaf may thus by its responsive movement indicate the passage of an excitatory impulse originated at a distance just as the excitation transmitted through a nerve is detected by the mechanical response of the attached muscle. In connection with this it should be remembered that the plant indicator may become inefficient under unfavourable circumstances. Absorption of excess of water, for example, will annul the motility of the pulvinus of *Mimosa.* Under these circumstances, we may have an excitatory impulse without any external manifestation of its passage through the plant. It will thus be understood how an excitatory impulse in an ordinary plant may pass unnoticed on account of the absence of an efficient motile indicator. The existence of such an impulse can, however, be detected by means of electric response.

In the present work the various excitatory phenomena of the plant have been investigated by means of mechanical response under the action of a testing stimulus. The different responses of plants may be included under three classifications: (1) *Simple response,* where a single stimulus evokes a single response; (2) *Multiple response,* where a single strong stimulus gives rise to multiple series of responses; and (3) *Automatic response* so called, where the pulsations appear to be spontaneous. In reviewing these in their proper sequence we shall be struck by the extraordinary similarities which are revealed between the response of the plant and the animal.

The Response Recorder

In the course of this work it has been shown that physiological changes induced in the plant owing to the action of the environment may be revealed by the characteristic variation of the amplitude and time relations of the normal curve of response. In obtaining records of response errors are, however, introduced on account of friction of the writing point against the recording surface. In recording the pulsations of *Desmodium* leaflet it is found that a weight as small as ·03 gram is enough to arrest the pulsatory movement. The difficulty of friction has been overcome by the method of intermittent instead of continuous contact for record. In the Resonant Recorder the writer is made to vibrate to and fro at a known and definite rate. The record consists of series of dots giving definite time intervals. In this manner time interval shorter than a hundredth part of a second can be measured. Owing to the extreme lightness of the recorder the error due to inertia is reduced to a minimum (p. 14).

METHODS OF STIMULATION

The various stimuli which evoke motile response in the animal are also found effective in giving rise to excitatory response in the plant. Thus the plant may be excited by mechanical, chemical, thermal, and electrical modes of stimulation. Electric stimulation may be caused by the polar action of a constant current, by the discharge of a condenser, or by the induction current. The electrical method allows the intensity of stimulus to be maintained constant, or varied in a quantitative manner. As in the skeletal muscle of animal, so also in the pulvinus of *Mimosa*, the break induction shock is more effective than the make shock. The sensitiveness of *Mimosa* to an electric shock is very great. It often reacts to an intensity of shock which is only one-tenth of the minimum perceived by a human subject (p. 23).

TIME RELATIONS OF THE RESPONSIVE MOVEMENT

Different plants exhibit different characteristics of response. The reactions are relatively quick in some and slow in others. In a typical case of *Mimosa* in summer the latent period was one-tenth of a second. The maximum fall of the leaf was attained in three seconds, and the recovery completed in 15 minutes. The rate of recovery was rapid at the beginning and very slow towards the end. The maximum rate of recovery was ·09 mm. per second in contrast to the maximum rate of fall of 24 mm. per second. The movement of recovery was about three hundred times slower than the movement of the excitatory fall. The extent of responsive fall in *Mimosa* increases with the increasing intensity of stimulus. The rate of movement is also increased under the action of stronger stimulus and higher temperature; it is decreased under fatigue. A stronger stimulus, generally speaking, requires a longer period for recovery. Under the physiological depression induced by winter the responsive reactions are modified, the latent period prolonged, and the amplitude reduced.

Biophytum sensitivum may be taken as typical of a quickly reacting plant. The leaflets undergo closure within a second after receiving the excitatory shock; recovery is accomplished in the course of three minutes. In marked contrast with this is the extremely sluggish reaction of the leaf of *Neptunia oleracea*, where the apex time is reached only after an interval of 180 seconds, and recovery completed in 60 minutes (p. 44).

Additive Effect of Stimulus

In the response of animal tissue it is found that a single stimulus by itself ineffective becomes effective on repetition. The same is found to be the case with plant tissues. Thus, with a given specimen of *Mimosa*, it was found that, while an electrical stimulus of intensity ·1 was singly ineffective, it became effective after being repeated twenty times. It is found, moreover, that this additive effect is, within limits, strictly quantitative. Thus with the identical specimen of *Mimosa*, when the intensity of individual stimulus was increased from ·1 to ·5, the number of repetitions necessary to cause effective excitation was reduced from 20 to 4. Here $\cdot 1 \times 20 = \cdot 5 \times 4$. The effective stimulation is thus found constant, being equal to individual intensity multiplied by the number of repetitions (p. 54).

Effect of Temperature and of Intensity of Stimulus

The response of *Mimosa* is abolished at a low temperature. With the rising temperature the amplitude of response is increased and the period of recovery shortened.

As in the response of animal muscle, so also in *Mimosa* there is a range between minimal and maximal stimulations where increasing amplitude of response occurs under increasing intensity of stimulus. The range within which the increasing effect is observed to take place is relatively extended in the case of plants in a somewhat sub-tonic condition. The range of variation is, however, restricted in specimens which are highly excitable (p. 61).

Work Performed by the Plant

During the responsive movement the pulvinus of *Mimosa* can do the work of lifting a weight. The effect of load on the response of *Mimosa* is similar to that on the contractile response of muscle. In both, under increasing load, the height of response undergoes a progressive diminution with shortening of the period of recovery. Within limits, the amount of work performed by the muscle increases with the load. The same is true of the work performed by the pulvinus of *Mimosa*. Thus it is found that, while under a load of 100 mg. the work performed was 1340 mm. mg., it became enhanced to 8666 mm. mg. under a load of 2000 mg.

The rate of work was found to be 7480 mm. mg. per second (p. 57).

Various Types of Response

The response of the pulvinus of *Mimosa* exhibits characteristics which are similar to those of the response of muscle. Under normal conditions, and with sufficient intervening periods of rest, the successive responses are found to be of uniform height. Under conditions of incomplete recovery, however, the responses exhibit signs of fatigue. The excitability of the plant in a sub-tonic condition is enhanced by the action of stimulus itself. Under such conditions the response exhibits a staircase increase.

Under extreme sub-tonicity the normal negative response may even be converted into abnormal positive.

The anomalous erection, after a preliminary fall of the leaf of *Mimosa* under continuous stimulation, is explicable on the common characteristics of response in plant and animal tissue. In both contraction is reversed to relaxation under fatigue (p. 84).

Variation of Motile Excitability under Changes of External Condition

The motile excitability of *Mimosa* undergoes abolition under sudden change from light to darkness. After a certain time the plant accommodates itself to the changed condition, with the restoration of normal excitability.

There is again a variation of motile excitability depending on the time of the day. The motility is at its maximum at a certain hour, and minimum at a different hour. This peculiarity is probably connected with the question of variation of turgor. The excess of turgor in the pulvinus caused by absorption of water is attended by a reduction or abolition of motile response. The lost motility may, however, be restored by the application of glycerin.

Very characteristic are the effects exerted by the different gases. Some induce a stimulating action, others give rise to depressing or toxic effects. Ozone enhances the excitability. Carbonic-acid gas and vapour of alcohol induce a moderate depression of excitability from which the plant recovers on readmission of fresh air. Coal gas and vapour of carbon disulphide also induce a depressing effect. The vapour of ether exerts a moderate narcotic action. The effect of chloroform is far more pronounced, the loss of excitability under its action being more complete and persistent. Ammonia has marked effect in the abolition of excitability. Sulphuretted hydrogen, nitrogen dioxide, and sulphur dioxide are very toxic in their action; their application is attended by quick abolition of excitability followed by the death of the plant (p. 94).

Death-spasm in Plant

One test by which a dead plant may be distinguished from a living one is that of electric response. The response of galvanometric negativity characteristic of living condition is abolished at death. When the plant is subjected for a time to a temperature of 60° C. its electric response, generally speaking, disappears. This temperature is, therefore, fatal for most plants.

When the leaf of *Mimosa* is continuously raised in temperature there is produced a progressive erectile movement; but at a critical temperature the erectile movement is suddenly reversed into a spasmodic contraction. This inversion takes place under standard conditions at or about 60° C., after this the response of the plant is permanently abolished. The death record is a V-shaped curve, the point of inversion being the death-point. After death a repetition of experiment shows no further inversion. Various other plants, sensitive and ordinary, exhibit this characteristic death-spasm at or about 60° C. In radial organ this movement consists of an abrupt longitudinal contraction. In taking an electrical record it is found that an electric-spasm also takes place at the critical temperature which is very near 60° C.

The death contraction in plants is similar to that seen in the animal. The death-point is found lowered under physiological depression. Thus, in a certain case, fatigue lowered the death-point of the plant from the normal 60° C. to 37° C. In another case dilute solution of poison lowered the death-point by 18° C. (p. 106).

Polar Effects of Electrical Current in Excitation

The fundamental unity of excitatory phenomena in the animal and plant finds a striking illustration in the characteristic effects induced at the kathode and anode.

In animal tissues, with feeble current it is found that excitation takes place only at the kathode at make. On moderately increasing the current excitation is found to take place at the make of kathode and break of anode. These effects are included in Pfluger's law of polar excitation in animal tissues. I have shown that effects precisely similar to these take place in the vegetal tissues. That is to say, the laws of polar excitation in plant under feeble and moderate currents are:

I. With feeble current the kathode excites at make and not at break. The anode excites at neither make nor break.

II. With moderately strong current, the kathode excites at make and not at break. The anode excites at break and not at make.

The polar reactions in the undifferentiated protoplasm of the plant body are thus identical with those of highly differentiated animal tissues (p. 233).

The effect of feeble ascending and descending currents in the petiole of *Mimosa* are parallel to those in nerve-and-muscle preparations. In both cases excitation takes place earlier when the kathode is nearer to the responding organ. As in animal, so also in *Mimosa*, single induction shock of moderate intensity is, as regards polar actions, effective at the commencement and not termination of the current (p. 206).

The sensitiveness of *Biophytum* to an electrical current is remarkably high; compared to the very sensitive human tongue, the sensitiveness of *Biophytum* is about ten times as great (p. 251).

Pfluger's law cannot be taken as a complete statement of the polar action of currents (p. 265). For strong currents there are induced two additional types of reaction:

III. Under the action of strong current, excitation takes place at the make of kathode and make and break of anode.

IV. Under still stronger currents, excitation takes

place at the make and break of both kathode and anode.

Physiological changes are also found to modify the polar effects of current. Under the separate or joint action of an increasing current and physiological modification, types of reaction are observed in plants which are similar to the so-called anomalous response in *Protozoa* (p. 276).

Contrasted Effects of Anode and Kathode

In the animal heart the contrasted effects of anode and kathode are exhibited by characteristic modification of its pulsating activity. Effects precisely similar are shown in the rhythmic tissue of *Desmodium gyrans*. The effect of the make of anode is to induce an expansion ; this is shown in the record of pulsation of *Desmodium* by a reduction of normal limit of systolic contraction. The opposite effect of contraction due to the make of kathode is seen in the reduction of normal limit of diastolic expansion (p. 236).

In nerve-and-muscle preparation the effects of ascending and descending currents are found to be modified by the intensity of the current. Effects in every way parallel are observed in experimenting with petiole-pulvinus of *Biophytum*. These characteristic modifications are easily traceable in *Biophytum* to the contrasted effects of anode and kathode and of make and break. Excitability is enhanced by the make of kathode and break of anode. It is depressed by the make of anode and break of kathode (p. 239).

A very important problem in plant physiology relates to the question as to whether in plants there is any transmission of true excitation. The prevailing opinion has been that in plants like *Mimosa pudica* there is merely a passage of hydro-mechanical disturbance, unlike the transmission of excitatory protoplasmic change, which takes place in an animal nerve. This view has been based on the experiments of Pfeffer and Haberlandt, who found transmission of stimulus to take place in spite of narcotisation or scalding of the intervening tissue. It is shown that those experiments are not conclusive, inasmuch as superficial narcotisation or scalding is not effective in abolishing the conducting power in the interior of the tissue.

For the settling of the question whether the transmission is physical or physiological it was necessary to have quantitative measurements of the highest accuracy in order to determine whether physiological changes affect in a definite manner the velocity of transmission. For the accurate determination of velocity it is essential to allow for the latent period of the responding pulvinus and its variations under different conditions.

Determination of Latent Period

By the employment of the Resonant Recorder the value of latent period can be accurately determined within a hundredth part of a second. Successive determinations of the latent period under constant external conditions are found to give identical results. The shortest value of the latent period in vigorous *Mimosa* is ·06 second, the average value in summer being ·1 second. The latent period is in general shorter under stronger intensity of stimulus; but the value becomes constant above a maximal stimulus. In the optimum condition of the plant the latent period is the same for feeble or strong stimulus. Fatigue prolongs the latent period; a rise of temperature, on the other hand, shortens the latent period (p. 130). The latent period of the pulvinus of *Neptunia oleracea* is ·6 second.

Determination of Velocity of Transmission of Excitation

Successive values of velocity of transmission are found constant when the applied stimulus is constant, and when the intervening period of rest allows complete protoplasmic recovery. Consistent results have been obtained by the employment of the *Direct* and the *Differential Methods*. The automatic records afford measurement of time as short as ·05 second. The highest velocity of transmission of excitation that has been found in the petiole of *Mimosa* is 30 mm. per second. In a sub-tonic tissue the velocity of transmission of excitation is enhanced under increased intensity of stimulus. The tissue became a better conductor of excitation in consequence of previous stimulation. Fatigue depresses the rate of conduction of excitation. Velocity of transmission becomes markedly enhanced at a higher temperature (p. 149).

The transmission of excitation takes place in both directions; but the velocity is not necessarily the same in the two directions. In *Biophytum* the velocity in the centrifugal direction is quicker than in the centripetal.

Excitatory Character of the Transmitted Impulse

The enhancement of velocity under favourable condition of rising temperature proves the physiological character of the transmitted effect. In support of this there are various confirmatory proofs, some of which may be regarded as crucial.

Transmission of excitatory electric impulse.—The excitatory change in the plant is accompanied by a concomitant electric change of galvanometric negativity. Electrical investigation shows that this excitatory electric impulse is transmitted to a distance through certain plant organs.

Excitatory impulse in absence of mechanical disturbance.—The hydro-mechanical theory presupposes the occurrence of a strong mechanical disturbance to give rise to the transmitted impulse. But initiation of excitatory impulse is found to take place under the polar action of an electrical current in the absence of any mechanical disturbance. This is realised when we find that an excitatory impulse is initiated and transmitted by the action of a current which is so feeble as not to be perceived by the very sensitive human tongue. The further fact that the excitation occurs only at the point of kathode at make and at anode at break shows that the effect transmitted is not physical but physiological.

Multiple excitation by constant current.—According to the mechanical theory multiple excitation can only occur under separate hydro-mechanical disturbances caused by multiple blows. It is, however, found that as in the rhythmic tissue of the animal, so also in that of the plant, multiple excitations are induced by the action of a constant current (p. 249).

Characteristic effect of temperature on polar excitation.—Excitation of animal nerve by induction shock is enhanced by warmth and depressed by cold. The reverse is the case when the stimulation is caused by constant current. The excitatory effect here is depressed by warmth and exalted by cold. These specific effects are found repeated in the conducting petiole of *Mimosa*. The excitation caused by induction shock is depressed by cold and enhanced by warmth. But as in the animal nerve, so also in the petiole of *Mimosa*, these effects are reversed in excitation under the polar action of a constant current. The excitation is now enhanced by cold and depressed by warmth. Minimal excitation becomes maximal under cold and ineffective under warmth (p. 247).

The crucial test of the excitatory character of the transmitted impulse is furnished by the action of various physiological blocks which arrest the transmission of excitation.

Paralysis of conduction by cold.—The local application of increasing cold on the conducting petiole retards and finally arrests conduction of excitation. As an after-effect of the application of cold the conducting power is paralysed for a considerable length of time. The lost conducting power may, however, be quickly restored by tetanising electric shocks (p. 164).

The electrotonic block.—The excitatory impulse may

also be arrested by the action of electrotonic block. This arrest persists during the continuation of the blocking current, the conductivity being restored on its cessation (p. 167).

Block of conduction by action of poison.—Finally, the conductivity of a selected portion of a petiole may be abolished by the local application of poison. The abolition of conducting power proceeds slowly under the action of copper sulphate solution and quickly under potassium cyanide (p. 173).

These results prove conclusively that the transmission of excitation in plant is a process fundamentally similar to that which takes place in the animal, being in the one case as in the other a propagation of protoplasmic change.

Direct and Indirect Effects of Stimulus

When stimulus is applied directly on the responding organ there is induced an excitatory fall of the leaf, concomitant with contraction and negative turgidity-variation. This particular reaction is designated as the Direct Effect of Stimulus. When, on the other hand, a feeble stimulus is applied at a distance there occurs only a positive or erectile response of leaf with concomitant expansion and positive turgidity-variation. This particular reaction is designated as the Indirect Effect of Stimulus. If the intervening tissue be highly conducting and the stimulus sufficiently strong, then the excitatory negative effect masks the positive. In such a case the response to indirect application of stimulus is negative—that is to say, the same as caused by direct stimulation. But, if the intervening tissue be semi-conducting, or if the stimulus be feeble or applied at too great a distance, then there is induced the positive or Indirect Effect (p. 196).

These two opposite reactions are found to take place in various plants under definite conditions and under diverse forms of stimuli. The two opposite effects are demonstrated independently by means of mechanical and electric responses.

Two diametrically opposite effects are thus induced by an identical stimulus, depending on direct or indirect application. The existence of the positive or the indirect effect of stimulus has hitherto been unsuspected. It must be taken into full consideration in unravelling the complexities of reaction in a responding organ.

The laws of DIRECT and INDIRECT EFFECTS OF STIMULUS are :

The effect at the responding region of a strong excitation transmitted through a short distance or through a good conducting channel, is negative, being the same as the effect under direct stimulation. The response is by negative turgidity-variation, contraction, fall of leaf, and electrical change of galvanometric negativity. This is the direct effect of stimulus.

The effect of feeble stimulus transmitted through a great distance or through a semi-conducting channel, is positive. The responsive reaction is by positive turgidity-variation, expansion, erection of leaf, and electrical change of galvanometric positivity. This is the indirect effect of stimulus.

Multiple Response

In taking records of electric response it is often found that, while a single moderate stimulus gives rise to a single response, a strong stimulus gives rise to a multiple series of responses. Similarly in *Biophytum* and *Averrhoa*, while a moderate stimulus gives rise to a single mechanical response, a strong stimulus gives rise to a multiple series of responses. These multiple responses are induced by various modes of strong stimulation, such as induction shock, constant current, strong light, thermal shock, and chemical excitation (p. 285).

When *Biophytum* is subjected to successive stimuli

of increasing intensity, the amplitudes of the successive responses remain the same. The response is, therefore, on 'all-or-none principle.' After the stimulus intensity has reached a certain limit the excess of absorbed energy finds expression in multiple responses. Certain plant tissues have thus the power of holding the excess of stimulus latent, to be given out later in the form of recurrent responses (p. 282).

The characteristics of the response of *Biophytum* are like those of cardiac tissue of the animal. Both are characterised by a long refractory period and response on 'all-or-none principle.' In both a single moderate stimulus gives rise to a single response and strong stimulus to a multiple series of responses.

There is no strict line of demarcation between the phenomenon of multiple and of spontaneous response so called. Under very favourable circumstances of absorption of excess of energy from without an ordinary responding plant like *Biophytum* will become converted to an automatically responding plant like *Desmodium gyrans.*

Automatic Pulsations of *Desmodium gyrans*

No satisfactory theory has been offered in explanation of the so-called spontaneous movement. It has, however, been shown in this and in my previous work that there is no such thing as an absolutely spontaneous movement, but that every movement is the result of the action of stimulus which has been stored up. That this is the case may be demonstrated in the case of *Desmodium* by isolating the leaflet from external sources of stimulation. The effect of run down of stored up energy is then seen in the gradual stoppage of the pulsatory movement. In this condition of standstill response occurs under fresh stimulation. If the depletion of energy has not been excessive, then a moderate stimulus gives rise to a multiple series of responses. But, under greater depletion, a strong stimulus evokes only a single response (p. 316).

Desmodium leaflets in a state of standstill give a single response to a single induction shock of moderate intensity. In a typical case the latent period was found to be ·4 second, the apex time 45 seconds, and the period of relaxation 120 seconds. The response curve exhibits a flattened top.

In summer a single pulsation of a vigorous leaflet of *Desmodium* is accomplished in the course of about 100 seconds. The quicker down movement is completed in 41 seconds, the maximum rate being ·9 mm. and average rate ·44 mm. per second. The period of up movement is slower, being 60 seconds; maximum rate of up movement is ·56 mm. per second, the average rate being ·3 mm. per second.

The pulsating activity of the detached leaflet of *Desmodium* can be maintained uniform for several hours by subjecting it to a moderate internal hydrostatic pressure. When the internal hydrostatic pressure is increased, the limit of diastolic or up movement is increased; the contractile movement being opposed, the systolic limit is decreased. Under increasing external load the pulsation is decreased in amplitude and is finally arrested (p. 301).

Similar Characteristics of Rhythmic Pulsation in Animal and Plant

The rhythmic tissues of the plant exhibit characteristics which are extraordinarily similar to those of the rhythmic tissue of the animal. The cardiac tissue of the animal has a long refractory period; the tissue takes no account of a stimulus which falls within the refractory period. This is also characteristic of the response of the rhythmic tissue of *Desmodium* (p. 310). The rhythmic tissues, animal and vegetal alike, are incapable of tetanus.

By the application of Stannius' ligature the pulsation of the heart is arrested at diastole. A similar arrest at diastole is found to take place in the pulsation of *Desmodium* by the application of ligature below the motile organ (p. 303).

The pulsating leaflet of *Desmodium*, like the pulsating heart, is more susceptible to excitation at diastole than at systole. An extra pulsation is induced by an induction shock applied during the diastolic phase (p. 319). Transmitted excitation affects the normal pulsations of rhythmic tissues—animal and vegetal—in a similar manner. In certain circumstances the effect is one of inhibition; in other circumstances the effect is one of acceleration (p. 320).

Still more remarkable are the similarities of effect of temperature and of chemical reagents on the rhythmic pulsations in animal and plant.

Effect of Temperature

The effect of lowering of temperature on the rhythmic pulsation of *Desmodium gyrans* is similar to that on the pulsation of a frog's heart. Lowering of temperature enhances the amplitude, but reduces the frequency of pulsation of both. The pulsation of *Desmodium* leaflet is arrested at the minimum temperature of about 17° C. Arrest takes place at systole; gradual warming revives the pulsation, which undergoes a staircase increase with enhancing diastolic expansion (p. 326).

Rise of temperature induces enhanced frequency and diminished amplitude of pulsation. During rise of temperature to about 43° C. there is a tendency of arrest towards diastole. The systolic contraction undergoes continuous diminution during rise of temperature. During the fall of temperature there is a gradual enhancement of systolic contraction (p. 330). The temperature maximum at which arrest of pulsation takes place may be as high as 45° C. Above this temperature there is a tendency to contraction and permanent arrest under heat-rigor.

Effect of Chemical Agents

The effect of drugs on the rhythmic pulsation is modified by the tonic condition of the plant, the strength of the reagent, and the duration of application.

Vapour of alcohol and dilute carbonic acid induce a transient enhancement of amplitude with prolongation of period. Stronger application induces an arrest of pulsation. Dilute vapour of ether and carbon disulphide induce a temporary arrest, revival taking place after quick substitution of fresh air. The action of chloroform is more intense than that of ether.

Copper-sulphate solution causes a permanent arrest of pulsation. The poisonous reaction of potassium cyanide is more powerful than that of copper sulphate.

A very striking characteristic modification in the rhythmic activity of animal tissue is found in the antagonistic action of acid and alkali on the pulsation. Application of dilute acid induces in the heart an atonic reaction with arrest of pulsation in the relaxed or diastolic condition. The effect of alkali is the very reverse of this, the arrest taking place in systole. These specific effects are reproduced in an astonishing manner in the rhythmic pulsation of *Desmodium.* Dilute solution of lactic acid induces in the pulsating leaflet an arrest at diastolic relaxation. The application of dilute sodium hydrate induces, on the other hand, exactly the opposite effect of arrest at systole (p. 339).

At the beginning of this work we took up the question of the possibility of detecting internal changes in a plant by subjecting it to a questioning shock. It has been shown how the plant can be made to record its answer to an impinging testing stimulus, and how the effects of environmental changes may be read in the script made by the plant itself.

It has been shown that the variations in the plant's, physiological activity, under changing external conditions may be gauged by the waxing or waning of its response. It has been shown also how numerous and varied are the factors that go to make up the complexity of the responses in the plant. It has been shown that stimulus may be modified in its effect, according as it is direct or indirect, according as it is feeble or strong. The modifying influence of the tonic condition of the tissue has also been shown, depending on whether it was normal, sub-tonic, or fatigued. In the numberless permutations and combinations of these varied factors lies the infinite complexity of the responsive phenomena of life.

In surveying the response of living tissues we find that there is hardly any phenomenon of irritability observed in the animal which is not also found in the plant. The various manifestations of irritability in the plant have been shown to be identical with those in the animal. From the standpoint of the theory of evolution this will be found highly significant. It may be confidently predicted that the recognition of this unity of response in animal and plant will in no small degree further the progress of plant physiology. Many difficult problems in animal physiology, moreover, will find their solution in the experimental study of corresponding problems under simpler conditions of vegetable life. The study of the responsive reactions in plants must, therefore, be regarded as of fundamental importance in the elucidation of various phenomena relating to the irritability of living tissues.

CLASSIFIED LIST OF EXPERIMENTS

Simple Mechanical Response

Different Types of Response

Modification of Polar Reaction under Tissue Modification

Characteristics of Polar Excitation

Contrasted Effects of Anode and Kathode

Multiple Response

Automatic Pulsation of Desmodium

Condition of Standstill and Revival of Pulsation

Effect of Electric Stimulation on Automatic Pulsation of Desmodium

Influence of Temperature on Rhythmic Pulsation of Desmodium

Action of Chemical Agents on Rhythmic Pulsation of Desmodium

INDEX